CAVES AND CAVING IN BRITAIN

Caves and Caving in Britain

EDMUND J. MASON

ROBERT HALE . LONDON

© *Edmund J. Mason 1977*
First published in Great Britain 1977

ISBN 0 7091 6195 6

Robert Hale Ltd.
Clerkenwell House
Clerkenwell Green
London EC1

Printed in Great Britain by Billing & Sons Limited,
Guildford, London and Worcester

Contents

Illustrations

Caves and Caving in Britain

Between pages 112 and 113

MAPS

PICTURE CREDITS

Peter Anthony, 32; Betty J. Barrett, 27; John H. Barrett, 1, 14, 15, 35, 37; B.O.C. Ltd, 8; The British Travel Association, 29; Dr Alan C. Coase, 4, 6, 9, 10, 12, 28, 30; Dan yr Ogof Caves Ltd, 5; Paul R. Deakin, 3, 18, 19, 20, 23, 24, 25, 31, 33, 36, 38; J.H.D. Hooper, 11, 13, 34; Edmund J. Mason, 7, 16; Stump Cross Caverns, 22; Treak Cliff Cavern, 26; Wookey Hole Caves Ltd, 17; J.R. Wooldridge, 2, 21.

Acknowledgements

I thank the following for assistance with this book: Mrs Dorrien Mason, who spent many hours in research, typing and correcting the script and without whose efforts the book could not have been completed; and the photographers and companies who supplied photographs, namely Peter Anthony, Mrs Betty J. Barrett, John H. Barrett, B.O.C. Ltd, The British Travel Association, Dr Alan C. Coase, Dan yr Ogof Caves Ltd, Paul R. Deakin, J.H.D. Hooper, Stump Cross Caverns, Treak Cliff Cavern, Wookey Hole Caves Ltd, J.R. Wooldridge and the Mendip Nature Research Committee whose copyrights are acknowledged.

My thanks are also due to Anthony Bagshaw and Andrew J. Barker of White Scar Caves Ltd, Mr and Mrs G. Gill of Stump Cross Caverns, C.L. Holcorn and H.J. Pugh of the Great Rutland and Masson Caves, C.G. Hunt of Wookey Hole Caves Ltd, Ashford Price of Dan yr Ogof Caves Ltd, Mrs Clarke, John Letheron and C. Richards.

I am also indebted to the many cave club journals and other publications and their writers, particularly in connection with some of the specialist branches of speleology, and particularly to the *British Caver,* formerly edited by Gerald Platten and later by Tony and Anne Oldham, *Studies in Speleology* by the Association of the Pengelly Cave Research Centre, the Transactions of the Cave Research Group of Great Britain, and the Bulletins and Transactions of the British Cave Research Association.

Lastly, my thanks are due to all those who have given me assistance and to whom, unfortunately, space does not allow individual thanks.

Introduction

IN A WAY, one might say that this book is a justification for my own caving days. Many people have asked me why I went caving. Firstly I must admit it was because I liked to cave. It was a challenge and an adventure and I appreciated the companionship of the caving fraternity, but I hope I have shown in this book that there is a good deal more in caving than mere adventure.

This book is not intended as a comprehensive work on caves or caving. The subject is too vast for a book of this size. It is essentially an introduction to a subject which has become popular over the years. When I first went caving, there were very few cave clubs and not many cavers. Those who followed the sport for any length of time, or who carried out research in caves, were known to each other, at least by name, whether they were operating in Yorkshire. Derbyshire or Mendip. Today there are many clubs and a very great many caves. Caving is one of the sporting school projects and some schools have their own clubs, as have many large industrial and commercial undertakings.

In spite of all this activity, the majority of people know little of caves and cavers, beyond the occasional news flash of an accident or dramatic rescue. Because of this, many believe that caving is just a dangerous sport and, when the rare accident does occur, clamour for caves to be closed. I hope that this book will do something to bring caves and caving into true perspective as a responsible interest and one that promotes team spirit and comradeship.

To the ignorant or careless, caving can be a dangerous occupation, but the dangers are much reduced when certain rules are followed. Anybody who contemplates caving, for his own sake as well as that of his companions, should join a recognized club and cave with people who know the caves of the area. Nobody, of course, should ever cave alone or indeed without an experienced caver in the party. There are many advantages in joining a club, for not only are you likely to become a 'safe' caver and earn the respect of your caving colleagues, but a whole range of activities becomes open to you, new exploration, cave photography, hydrology and many other caving subjects. Even if you are only attracted by the sporting aspect, a club can offer many facilities, the use of tackle and usually a club hut.

This book is intended not only for the beginner in caving, but also for the reader who wants to know what caving is all about, and it only deals with the subject in a general manner. Those who want to pursue any of the aspects further must turn to more specialized works and for this reason an extensive bibliography has been included, although, even then, it has not been possible to include every book on the subject.

There has been no attempt to list and describe all the important caves which exist in each region. This has been adequately done in various individual publications and would be tedious in a book of general interest. Instead, I have described in some detail the exploration of certain caves only and left the rest to the bibliography.

The caving areas of Britain cover a large geographical field. New discoveries are made constantly and it is not possible to keep up with all that occurs in every region. Being a southerner myself, most of my caving was done in Mendip and in Wales and there is a tendency to deal in greater detail with those sites I know best. I hope the northern potholer will forgive me if he or she feels that I have not given adequate space to the happenings in the region nearest their own interests.

To the casual reader, I hope this book will bring a greater understanding of those who constantly explore below the surface of Britain, and to the caver or would-be caver, may I use the caver's greeting of "Good caving!"

ONE

Cave Country

MOST PEOPLE'S EXPERIENCE of the world of caves begins with a visit to a show cave, although these are but a small proportion of known caves. Visitors to show caves may notice that, however far apart are these show caves. there are similarities in the country surrounding them. Such caves are usually on the edge of hill or mountain country, ridged with deep gullies and gorges, such as Cheddar and Wookey Hole caves at the foot of the Mendip Hills, Dan yr Ogof cave in its ravine beneath the Cribarth, Peak Cavern under the moor at Castleton, and Ingleborough Cave on the edge of Ingleborough Moors. Through these caves flow underground rivers to emerge in daylight at or near their entrances and these rivers have formed spacious chambers which make these caves suitable for tourists.

There are countless other caves to be found high in the moorlands, where the water enters the cave systems and where, depressions and outcrops of limestone are characteristic of cave country. Here is a country of sink holes and potholes, of deep depressions where streams go underground through the cave systems. Except where there are pockets of water-holding clay and peat, the mountain plateau is hard and dry, partly covered by a thin mat of poor soil, carrying springy turf and small limestone plants. Here and there, hard, grey-white buttresses of rock stand out against the sky and on steep slopes are unstable screes and everywhere rocky outcrops.

There are four main cave areas of Carboniferous Limestone: Mendip, South Wales, Derbyshire, and most important of all, the Northern district. There is a smaller area of caves in Devon in Devonian Limestone.

The first range of hills you see, after leaving Bristol for the south-west, is the Mendip Hills, more familiarly known to

CAVE AREAS OF BRITAIN

local people as Mendip. They extend from Frome to the sea at Weston-super-Mare. Their height is magnified by the fact that they rise from undulating but comparatively low country, for the highest land of the range, the long, slightly convex, Old Red Sandstone cap of Blackdown, is only just over a thousand feet above sea level.

To the casual visitor, Mendip means Cheddar Gorge, the home of cheese, strawberries, clotted cream and the Cheddar caves. It is, of course, only a small part of Mendip, an impressive rift in the side of that Carboniferous Limestone mass which makes up for the bulk of the hills. There are limestone gorges elsewhere in Britain, but Cheddar Gorge is spectacular because the narrow defile brings the rocky walls close together, until the high pinnacles stretching skywards seem to threaten to tumble down on the visitor at any moment.

Popular as the gorge may be, to the caver, it is only one of several caving areas on Mendip. As he travels from Bristol towards Weston-super-Mare, he sees the wooded gap which is Burrington Combe, a gentler gorge than Cheddar and where, almost opposite the famous Rock of Ages, is Aveline's Hole, an archaeological rather than an exploratory cave. Higher up the combe, the Twin Brook valleys can be seen, making their way up to Blackdown. The Lower Twin Brook valley contains Goatchurch Cave with its intricate system at different levels. On the opposite side of this valley is Sidcot Cave, named after the boys of Sidcot School who found it—a small system noted for its tightness.

Beyond, on the edge of Blackdown, is Read's Cave, once known as Celtic Cavern because of the prehistoric Iron Age material found in the main chamber. Further along the range is Wavering Down, with its Coral Cave, where I once fell out of a bosun's chair and was suspended on a lifeline. Close by is the familiar landmark of Crook or Crook's Peak, the only peak in Mendip. In spite of its appearance of height, it falls far short of the height of Blackdown. Near Weston-super-Mare, we have outlying heights such as Banwell Hill, with its bone cave and stalactite cave, and at Weston the hills dip down to the sea at Brean Down and Sand Point, to reappear again as the two mid-channel islands of Steep Holm and Flat Holm.

The main resurgence caves of Mendip are the show caves of Cheddar and Wookey Hole, while the inlet caves where the water goes underground to feed these systems are principally in the uplands round Charterhouse and Priddy. Priddy Green, famous for its Priddy Fair, is a focal point for cavers.

There, on most weekends, can be seen groups of cars and minibuses, their helmeted passengers assembling on the green with ropes and equipment to set off for the nearby caves of Swildon's Hole and Eastwater Cavern. Priddy is the very heart of Mendip. It is a scattered village, more green than residential—a village which seems to have stood still through the years, with its dry-stone walling and the cluster of prehistoric burial mounds on the skyline. It is the hub of a network of narrow, undulating roads leading over the great back of Mendip to the caves.

One road leads down the steep hill to Wookey Hole, past the Nature Reserve of Ebbor Gorge. The limestone cliffs of the gorge can be seen through the trees to the left. To the right are fine views across the flats to Glastonbury Tor, where the last Abbot of Glastonbury was hung, drawn and quartered.

Another road from Priddy winds its way past numerous burial mounds, turns at the Miner's Arms, no longer a tavern of the lead miners, to the Castle of Comfort Inn. The road to the Harptrees takes you to Gibbet's Brow, close to the shaft of Lamb Leer, a large cave system in the ancient miners' 'gruffy' ground, the name given to land full of the hollows and hillocks of the old lead workings.

From Priddy, too, a road leading from the direct road to Cheddar passes near ancient burial mounds and more gruffy ground at Velvet Bottom to Charterhouse. You may see the church, but you will not find the village at Charterhouse. There isn't one, although remains of a large Roman settlement lie beneath the turf. The last of the lead resmelting community, who have left the old ruined works and debris heaps of black slag across the valley, have long since gone. Today the church serves the few scattered farms and houses in the area. Charterhouse is more the name of a district than a settlement and it is in this district that we find the Charterhouse system of caves, such as Longwood and August Hole and G.B. Cave. Finally, you can leave Charterhouse and Mendip through Burrington Combe.

Although the South Wales caving area has geographical affinities with the other three cave districts, it is different in some respects. It is a mountainous country with a language of its own. The names of its caves, like other place names, may present some difficulty to the average Englishman, but it takes a surprisingly small vocabulary to translate them into English and the interest this creates can add to the pleasures of a Welsh caving holiday.

The approach to the caving area from the South Wales coast is through the industrialized area of steel works and coal mining. Coming north from Swansea, the valley road offers nothing but factories, mines and rows of small houses. Even the sides of the mountains beyond the valley seem to have suffered the scars of industry. Then, as we cross the northern fringe of the coal belt, the sordid landscape disappears and there ahead, bathed in clean air and bright sunlight, if you are lucky, are the mountains of the caving country, rising higher and higher until they culminate eventually in the twin peaks of the Brecon Beacons.

As we enter this unspoilt region, we pass a side lane which passes over the River Tawe and makes its way up a long incline to the disused Penwyllt railway station. This station was provided at the request of one of the most famous of opera singers, Madam Adelina Patti, who lived further up the valley from 1897 until her death in 1912. The Penwyllt railway station was her link with the outside world and from it she travelled to London and other cities and received her guests to her mountain home. Adjoining the railway station is the deserted quarry village of Penwyllt, where the South Wales Caving Club has its headquarters, adjoining huge quarries which were a threat to Britain's largest cave system—Ogof Ffynnon Ddu. The cave extends under wild mountain country to Pwll Byfre, where the main stream enters among high uplands to emerge in the outlet of Ffynnon Ddu on the bank of the River Tawe in the valley below.

Returning to the Tawe valley, we soon pass the pseudo

castle of Craig-y-Nos (Rock of the Night) adjoining the road.
It was here that Madam Patti lived and died. She chose this
spot because the valley and the mountains reminded her of
Italy. Here we have the first indication in the valley that we
are in cave country. Opposite the entrance to the castle, now
a hospital with its grounds as a country park, water gushes
from the mountainside and flows over an artificial dam. This
is Hospital Cave, which once furnished power to supply the
castle with electricity, but diving operations have failed to
penetrate an extensive system. As the road rises above
Craig-y-Nos, car parks and buildings on the mountainside
mark the site of the only two show caves in Wales, Dan yr
Ogof and Cathedral Cave. Above Dan yr Ogof, out of sight
of the road, is Ogof yr Esgyrn (Cave of the Bones) and on the
high mountain moorland, above and behind the resurgence
cave, are numerous swallets and depressions, some of which
connect with the Dan yr Ogof system and act as feeders to
the River Llynfel flowing through it. From this moorland
there is a fine view up the valley, with the distinctive slope of
Fan Gihirych on one side, the escarpment of Carmarthen Fan
on the other and the rounded Cefn Cul in the middle, where
the Trecastle road separates from the main road to Senny-
bridge. Opposite, across the Tawe valley, we can see the
quarries of Penwyllt, and over the mountain above them and
out of sight lies the valley of the Little Neath.

Here are such caves as Pwll y Rhyd and further to the east
again is the Mellte valley with the great entrance and system
of Porth yr Ogof. This is an area of waterfalls and dry river
beds, where the rivers go underground. Deep wooded ravines
cut their way into the mountain moorland. Seen from the air,
they are like long tongues of green, penetrating into the
barren uplands, but they are cut so deeply and narrowly that
seen across the moorland, they appear as mere depressions,
giving little indication of their depth, thickly wooded sides
and the rivers tumbling away at the bottom over one
waterfall after another.

The head valleys of the Neath and Mellte, the Tawe valley
and adjoining mountains form one vast caving area, but there
is another important area to the east of the Beacons at
Llangattock, near Crickhowell. The caves are mostly concen-

trated in a small area along the old trackway at the base of the bluff above Llangattock. Most of them are small, but one, Agen Allwed, ranks with Ogof Ffynnon Ddu and Dan yr Ogof as one of the largest systems. The walk along the track is worthwhile in itself. Across the Usk valley we can see Gwernvale, the house where Everest lived who gave his name to the highest mountain in the world, while up the valley we can see the point of the famous Sugar Loaf at Abergavenny.

Like South Wales, Derbyshire's caving country is grouped round definite centres. The two principal ones are Matlock Bath and Castleton. Matlock Bath is an attractive little spa of white buildings, wooded slopes and pleasant parkland on the banks of the Derwent. It has two adjoining show caves under the same control, Rutland and the Great Masson. There was a third at Matlock Bath, the Cumberland Cavern, but it is no longer open to the public. In this area, we have that feature not peculiar but particular to Derbyshire of the combination of mines and caves, for there are probably more combinations of cave and mine in the Derbyshire area than in any other caving region. Opposite the Heights of Abraham, which conceal the two show caves, the limestone nature of the area is revealed in the high limestone crags which border the Derwent. Beyond the crags is High Tor, surmounted by a ruined castle, clearly seen from the Heights of Abraham. Riber Castle is not an ancient castle, although it is now in ruins; it was built by John Smedley, a successful textile industrialist.

From Matlock Bath we can drive for about twenty miles to Buxton, 'the mountain spa'. Much of it is over a thousand feet above sea level and the heights around it rise to more than 1,500 feet. It is a typical spa town, with its imposing crescent and typical spa buildings. There is nothing of a speleological nature to keep us there though Poole's Cavern, long closed, has now re-opened; but we should pause to look into the museum, as it contains material collected from the Derbyshire bone caves.

As we go from Buxton to Castleton, the country becomes wilder and higher and we see far more mountain moorland. We hesitate at Dove Holes, for the name must certainly imply caves. The name is derived from darf holes which in

local dialect means 'dwarf' holes. No doubt the name has a
fairy significance, but such caves as exist are fairly dwarf and
so we push on towards Castleton.

About a mile and a half beyond Sparrow Pit, a path leads
off to Eldon Hill, the top of which is 1,543 feet above sea
level. On its flank is Eldon Hole. At various distances on both
sides of the road from this point to Castleton are a number of
caves and potholes, including Oxlow Cavern, Giant's Hole
and Nettle Pot. Further along the road, we come to Winnat's
Pass, the old road down to Castleton, but today the main road
takes an easier route. However there is still a steady incline
down to Castleton past Mam Tor, the Shivering Mountain, so
called because of the constant movement of broken silica and
shale down its side.

Castleton is a pretty village and there are many visitors,
particularly on Castleton Garland Day, 29 May, unless the
date happens to fall on a Sunday. The 'King', who is reputed
to represent Charles II, is encased in a huge framework of
flowers and rides on horseback through the village, followed
by his 'Lady' and the procession. At the end of the ceremony,
the Garland is hoisted up to one of the pinnacles of the
church tower and remains there while it is reasonably fresh.
Overlooking the village and above the great opening of Peak
Cavern is Peveril Castle.

The village itself has some quaint buildings and byways.
The Douglas Museum, originally provided by an escapologist
and friend of the great Houdini, as can be expected has a fine
collection of locks, but it also has a display of minerals and
some fascinating miniature models of tools. Castleton can
certainly be called the village of caves, for apart from the
potholes on the heights round it, there are no less than four
show caves, the Peak Cavern itself and the combination of
caves and mines, the Speedwell Cavern, Treak Cliff Cavern
and Blue John Cavern. I do not know of any place in Britain
with so many show caves in so short a distance. Only two
miles from Castleton is Bradwell with Bagshawe Cave and
one or two smaller ones as well.

The Manifold Valley over the southern edge of the county
is not strictly Derbyshire but Staffordshire. For caving
purposes, it is included in the Derbyshire Peak caving area.

Here the Manifold flows between great limestone cliffs, dotted with caves. They do not go very deep and are mostly of an archaeological nature, while on Derbyshire's Nottinghamshire border are the Creswell Crags with a series of further archaeological caves.

By far the most widespread and important cave areas of Britain are found in the north in parts of North Yorkshire, Lancashire, Durham and Cumbria. These areas are divided into several groups and at least one full-scale and several small books have been written about them. In this book, I can only deal with the most well-known and typical area of North Yorkshire, that centred around the villages of Ingleton and Clapham.

Like many of the northern cave districts, this is an area of potholes. Vertical shafts, sometimes very deep, are the entrances to the cave systems. The most productive area for potholes is the Carboniferous Limestone lying between and around the heights of Whernside, Ingleborough and Penyghent. The cave systems swallow rainwater which comes down from the three great heights. The water flows from the rain-saturated rocks to disappear into the adjoining limestone. A typical example of this action is Gaping Gill, where Fell Beck flows over the surface of the rocks but when it reaches the limestone outcrop, it plunges into the depths of the limestone. As you stand near the edge of the funnel of Gaping Gill and look over the windswept moorland, everywhere can be seen the depressions made by this water drainage.

As we walk over the moor, we not only have to beware of natural shafts, but below us, where the crust is thin, we can hear the streams gurgling beneath. This is very wild country, open moorland, with the typical dry-stone walling of the limestone area, an undulating country, stretching in all directions to merge into the slopes of the heights. Walking is hard going, picking one's way, sometimes over rough pebbly ground interspersed with tufts of grass, standing up like miniature islands, or through deep rough grass, sinking now and again into hollows in the surface. There are few trees and those that do exist seem to have the habit of growing out of the edges of potholes.

This wild country contrasts with the wooded valley where Fell Beck emerges near Ingleborough Cave. You come down from the moorland and through the limestone gorge of Trow Gill to the cave on the lower ground and walk through the woods of the Ingleborough estate, landscaped and cultivated with its placid lake and shrubs, to the pretty village of Clapham. Water overflowing from the lake tumbles in a great fall not far from the church. There is a National Park information centre where you can get a lot of advice about caves, potholes and the surrounding countryside.

Ingleton is another centre for the caving country and is only a short distance from Clapham. It is the larger village and the general impression is of grey stone buildings and water, for, as in the Upper Neath valleys, there are some spectacular waterfalls. A pleasant walk takes you over and across a series of these falls on the Rivers Doe and Twiss.

From Ingleton you can follow the beck up Kingsdale from where the old turbary road will take you past a series of potholes, including Rowten Pot and Yordas Cave, once a show cave but now closed. If we take the Hawes Road from Ingleton, we shall pass White Scar show caves, and further to the north in the area of Hill Inn, a cavers' tavern at Chapel-le-Dale, are a number of other pots including Jingle Pot, Hurtle Pot and Douk caves.

We come now from the northern and largest cave area to the southern and smallest of the main cave areas of Britain. In Devon, the contrast between upland and lowland caves is not so apparent. The country is one of low hills and the caves appear only in the occasional outcrops of Devonian Limestone. The cave areas are farmlands, very different from the open moorland of other districts. The geology of the Devon cave areas is far more broken up than that of the four main cave districts.

By far the most well-known cave in Devon is Kent's Cavern at Torquay, for besides being a show cave of considerable size and interest, it lies on the outskirts of a holiday town, on the western side of the Ilsham Valley, under a limestone hill, now a residential district. It is famous for archaeological discoveries made by early cave excavators in the nineteenth century and for research done there into the formation of stalagmitic deposits.

Another bone cave of interest is Brixham Cave, also known as Windmill Hill Cave and Philp's Cavern. It lies under a residential part of Brixham and when I saw it some years ago, it was entered through a house and used for furniture storage, but the furniture has now been removed and the cave is open to the public. This is a most interesting archaeológical site and well worth a visit. It is a strange coincidence that these two important bone cave sites of Devon are in the midst of built-up areas, which is a most unusual setting for any cave.

Devon may be a minor cave area, but it is a very important one for cave conservation. Travelling north from Plymouth, you leave the dual carriageway at the sign for Buckfastleigh and follow the tourists who are making for Buckfast Abbey. Turning up the steep Russetts Lane, you soon come to a quarry with a pleasing complex of stone-built buildings. This is the Pengelly Cave Research Centre and in the quarry is a series of caves used for education about caves in the interests of cave conservation. The Centre is named after William Pengelly, a nineteenth-century pioneer in the excavation of Devon caves. One building is used as a museum, demonstrating cave studies and emphasizing the urgent need for the preservation of our caves. Another building is being converted into a library, lecture room and accommodation units. Here the interested visitor can learn all about caves, from archaeological excavations to caving techniques and the need to join a club with a good record of conduct and conservation.

Every serious caver soon becomes aware of the need for cave conservation. The popularity of the sport has increased so enormously since the 1940s that sheer weight of numbers presents a serious problem, in the same way that the number of people mountaineering, walking or just visiting the countryside can cause increasing deterioration in the natural beauty they so admire. In the 1930s, the sport was confined to a comparatively few eccentrics who appreciated the adventure of an unusual sport. A sandwich dropped occasionally, or a box of matches lost down a crack, made little difference. A broken straw stalactite was regrettable, but not serious. On a caving trip, it was unusual to meet another party. You were almost certain to have the cave to yourself. Now in some popular caves, there is an almost unending stream of cavers and the problems are apparent. Well-established clubs are aware of these difficulties and take steps to make sure their members are also aware of them and do their best to preserve our caves in their natural state.

Another threat to our caves lies in quarrying activities, which in some cases mean the destruction of some caves and the creation of unstable conditions in others, leading to their closure. Many quarry owners are most co-operative, report any caves found in the course of quarrying and do what they can to avoid damaging existing ones. In other cases, some considerable difficulties exist.

Another serious problem facing the caving world is that of access to caves. To some extent this problem has also arisen because of the growth of the caving population. Landowners, worried by the numbers of cavers and by the possibility of liability for accidents, have closed caves on their land to the caver. In many cases, clubs have taken on the responsibility of controlling access and allow only responsible members to enter these caves, sometimes only with a recognized leader. Although many show cave owners welcome exploration by established clubs, some find it inconvenient to have parties of cavers walking through the show caves to reach the non-tourist depths and no longer allow access.

Although cave clubs are independent groups, these are problems which it was clearly advisable to deal with on a national scale. In 1969, the National Caving Association came into being, a policy body dealing with cave conservation and access, the welfare of cavers and the continuance of caving. The Association maintains close contact with public bodies on behalf of and through four member regional councils, the Council of Northern Caving Clubs, the Council of Southern Caving Clubs, the Cambrian Caving Council and the Derbyshire Caving Association. The National Association has done some valuable work at national level, administering grants from the National Sports Council and bringing problems of cave conservation and access before public bodies, but there are fears in some clubs that the organization will become too centralized and too far removed from the interests of the average caver.

Another national body is the British Cave Research Association, a body formed in 1973 as the result of a merger between the Cave Research Group of Great Britain and the British Speleological Association. The British Speleological Association was set up in 1935 under the presidency of Sir Arthur Keith. It was a body for co-ordinating cave records and making the information available to the caving clubs. In the journal of the Association, *Caves and Caving,* now defunct, were published lists of caves in the main caving areas, and lists of cave finds and the museums in which they were deposited. The Association co-ordinated the collection and study of fauna from cave waters and supplied to cave clubs and individual cavers the necessary equipment for the purpose. It also organized in 1937, on behalf of the Ministry of Health, the investigation into the course of underground water, with a view to supplementing water supplies and tracing any possible cause of pollution.

In 1947, the Cave Research Group of Great Britain was formed with similar objectives, collecting information and issuing apparatus for research. They circulated a newsletter to members and affiliated clubs, as well as publishing transactions and special papers on new discoveries and research. Under the editorship of C.H.D. Cullingford, they produced a composite book on *British Caving* which soon became the textbook of the serious caver. In 1950, the Group published a grading system for classifying plans according to their degree of accuracy, based on the types of instruments used. Memory sketches were classed as grade 1 and theodolite surveys as grade 7. At the same time, they unified the grading of caves which some writers had already been using, classifying caves into grades from 'easy' to 'super-severe'. By standardizing the grading, a fairly accurate idea was available of the conditions with which a caver would have to contend in any particular cave.

Shortly before the merger in 1973, the Cave Research Group revised the system of gradings, which were then adopted by the new body, the British Cave Research Association, as B.C.R.A gradings.

Another national body is the Cave Rescue Council, which represents cave rescue organizations in each area. Near the entrance of every major cave is a notice giving telephone numbers to contact in the event of an accident. Like most other caving activities, cave rescue has made great technical advances, including special stretchers and tackle adapted for cave conditions. The rescue teams have saved many lives, both of cavers and of members of the public. There will of course always be unavoidable accidents in an adventurous sport such as caving, but many accidents would not happen if newcomers to caving would join, learn, and cave with a good club.

THE CAVING CODE

(based on code devised by Bill Little of South Wales Caving Club, published by Brecon Beacons National Park Committee)

1. Obtain the permission of the owner or tenant to enter any cave.
2. Learn as much as possible about a cave's characteristics beforehand and, to avoid being trapped by sudden floods, consult local weather reports before entering the cave.
3. Never go in a party of less than FOUR PERSONS: if one is hurt, one can stay with the injured person and two can go for help. Otherwise ALWAYS KEEP TOGETHER.
4. Leave a note with some reliable person, giving the names of those in your party, which cave is being visited, and the expected time of return. Report back to that person when you return.
5. Wear clothing adequate to keep warm when wet in temperature of 9°C. (48°F.). Boots and helmets are essential.
6. Take food on any trip likely to take more than two hours.
7. Ensure that lights are in good order and that two separate types of lighting are carried, together with ample spares, e.g. bulbs, batteries, carbide, candles and matches, in a WATERTIGHT container.
8. Use rope with a minimum strength of 2000lb for lifelining on all climbs and ladder pitches.
9. Do not dislodge loose stones and never throw them down a pothole—someone may be below.

10. Do not dirty or destroy any formations or mark walls. Leave no rubbish in a cave or at the entrance.

TWO

How Caves are Formed

CAVES OWE THEIR ORIGIN to water, not only the cave passages and chambers, but the stalagmitic formations they contain. Most sea caves are made by the wearing away of soft patches of rock due to the battering of the sea, and inland caves by the solvent action of water on certain rock. Some sea caves, particularly those in limestone areas, may be a combination of both. It is obvious that inland caves are formed only in rocks susceptible to the solvent action of water. These rocks are certain limestones, particularly Carboniferous Limestone, and as they outcrop only in certain parts of the country, such caves are confined to these districts of Britain, principally parts of North Yorkshire, Derbyshire, Somerset, Devon, Wales and Scotland and a few other areas. Because the rocks are similar, the surface features of these districts are similar, as already described.

There are two ways in which the solvent action of water on limestone occurs, vadose action which takes place from the surface and phreatic action which happens below the water table, the natural level of ground water below the surface. Below this level, all underground passages would be water-logged until such time as there is a lowering of the water table.

In vadose action, rainwater, charged with carbon dioxide from the atmosphere and vegetation, drains down rock fissures in the surface of the ground and has a solvent effect on the limestone, so that the fissures become gradually enlarged. This process is helped by the abrasive action of loose stones carried along the enlarged cracks by the water. In fact, the action is twofold, solvent and abrasive. Below the surface, the water follows the fissures and it is these which become enlarged, eventually to form the cave passages.

The limestone was laid down by an ancient sea and forms a

series of beds. Cave water penetrating between these beds forms wide, low passages, usually tapering off at the sides and known as 'bedding planes'. Where the water eats down at the centre, a rift is formed along the floor of the bedding plane which accounts for many cave passages having deep recesses on both sides where they join the ceiling.

Although the beds were originally laid down horizontally, subsequent earth movement has tilted the beds so that they are now rarely on a level plane. As the degree of tilting varies from area to area, the general run of cave systems may have little gradient in one district and descend at a considerable incline in another, although one would expect several caves in the same vicinity to conform to a similar pattern, unless there had been extensive faulting or other disturbance to the rock beds.

As well as having the feature of the fission between the bedding planes, limestone breaks up into blocks because of vertical contraction cracks. These 'joints' run parallel, crossing others more or less at right angles, but again, although originally vertically at right angles with the bedding planes, this jointing is correspondingly tilted as are the bedding planes. It is along these joints that the water first makes its way into the system.

If the joints are reasonably vertical, the cave will be entered by a pothole, typical of the Yorkshire caverns, but in Mendip the entrance passages usually descend at a comfortable angle, and in Wales, such systems round the Little Neath and Swansea Valley can be almost flat planes. The term 'potholers' means just that in Yorkshire, but could not describe the cave explorers in southern areas, who call themselves just 'cavers', rather than 'potholers', since vertical entrance shafts are common in Yorkshire, but rare in the south.

Cave systems are usually fed from more than one source and when a second stream joins the main underground water, the limestone blocks at the junction tend to break down, forming a chamber. Such chambers are often strewn with large boulders because of the collapse of the rock.

Because of rock jointing, underground streams will often drop from one level to another, before meeting the water table or emerging as a resurgence, so that it is quite usual for caves to consist of a series of passages at various levels, linked by drops known as 'pitches', which may be shallow or deep. Obviously the passages at the lower part of the system would be more recent than those at the upper part, when they are formed by the same stream.

Cave systems are often cut across by river valleys and discharge their water into the rivers either directly or as tributaries. As the river cuts its valley deeper and the water table drops, the internal stream may make a lower underground course to join the river at its lower level and so leave the old resurgence passage dry and high above the river bank. It is not uncommon to find, in the cliff above the resurgence, a number of dry exits at different levels. The resurgence and the water-bearing passages behind it are known as the 'living' cave to distinguish the active system from the abandoned dry passages. Examples of this can be seen at Wookey Hole caves, where the Great Cave is entered by the dry system, while visitors can still see the river emerging from the arch in the valley below. Inside the cave, the visitor can descend to the river level or ascend by steps to the grotto which is part of an older level. An entrance to this older level can be seen in the cliff above the cave door. Similarly, at Dan yr Ogof caves, the visitor enters the dry system by an artificial tunnel and before doing so, can look down into the little valley to see the River Llynfel flowing through a rock arch from the living cave. At a higher level in the cliff face are older resurgences, including Ogof yr Esgyrn (Cave of the Bones), where a stream still flows in wet weather, but into a hole through the boulders within the cave to merge with the living cave system below. As the name 'Cave of the Bones' implies, such remnants of older dry systems found on cliff faces such as in the Manifold valley, Staffordshire, and in Cheddar cliffs, Somerset, have been used by men as habitations from the earliest times.

It seems possible that the small caves and rock shelters which sometimes appear on both sides of gorges, such as Ebbor Rocks, Somerset, are remnants of the side passages of a cave system and that the gorge or valley is the main cave with the roof collapsed. This theory has been suggested for the origin of Cheddar Gorge, but it is not universally accepted.

There is a theory that most caves owe their beginnings to phreatic action below the water table and that vadose action only came into play when the water table was lowered, leaving a cavity. The reason is the same, that the rock is soluble when in contact with water containing carbon dioxide, but with phreatic action, the dissolving process took place when the cave system was beneath the water table which has since lowered. The phreatic theories, and there are more than one, explain certain features in caves which are not cónsistent with vadose formation. There are variations in the scalloping, a well-known feature on cave walls, while certain tubular like tunnels are thought to be of phreatic origin. In a number of instances, both processes appear to have been at work, although at different times, but whatever process formed a particular cave, the basic feature is that caves are formed by rock solution in water.

The course of underground streams may change because of imperfections, such as faults, in the rock structure. Faults occur where earth movement in the past has been such that the rock deposits are cracked across, sometimes to such an extent that one or both sides of the crack has moved up or down, breaking the continuity of the beds. These are more extensive than mere jointing. Faults may be of considerable distance in length although they cannot always be recognized on the surface. In these circumstances, an underground stream may follow the fault rather than the jointing.

As limestone is a sea bed deposit, it contains sea fossils and these are often more easily studied in caves than in surface rocks. The fossils are harder than the cave walls and so stand out in relief against the water-worn rock. Carboniferous Limestone was laid down over a very long period and so can attain considerable depth and, of course, the fossils change according to the depth. Geologists zone the rock by the fossils it contains, naming the zones after the predominating types of fossils. As fossils usually have unwieldy names, initials are used. It would be tedious to keep referring to the Dibuno-phyllum Zone, so it is normally called the D Zone, which is itself divided into sub-zones D1, D2 and D3. D1 is the oldest and therefore lowest in the rock sequence. Caves are not confined to any particular zone.

The cave fossil most easily recognized by most people, and often seen in surface quarries, is the crinoid or sea lily. It has the appearance of small segmented tubes, usually masses of them jumbled together. Often only the cross-sections are visible as small discs in the rock. They look like plant stems. They were, in fact, not plants but sea animals related to starfish and sea urchins, although they have every appearance of plants even to their cup-like tops and were even fixed by their stems to the sea floor. They are known to many people as 'St Cuthbert's beads', as their jointed stems appear like segmented beads, while 'screw stones' is a common name for them. Some limestones are so full of them, as in parts of Derbyshire and Wales, that the rock has been given the false name of 'Crinoidal Marble'.

Other fossils include that of a type we call 'lamp shells', technically brachiopods. We are apt to call all shellfish 'molluscs' but lamp shells are not included in that category. Although they have an upper and lower shell (valve), they do not match as in most shellfish. One of the valves is beaked over the other and has a hole through which a 'foot' projects in life for attachment to sea plants. These brachiopods are known as lamp shells because of their resemblance to the oil lamps of Roman and Greek times on which the Toc H lamp is based. I have such a lamp from Carthage and, with the hole for the wick, it always reminds me of the brachiopods.

Carboniferous Limestone is noted for its coral fossils and in some areas the rock is known as Coral Limestone because of these. A particularly fine specimen of fossil coral projected from the roof of a passage in Sidcot Swallet in Burrington Combe many years ago but it has long since disappeared, a victim of fossil hunters.

The name Mountain Limestone is still often used, an old term which gave way to the more technically correct Carboniferous Limestone because of association with the coal measures. As a limestone it is akin to many other rocks in Britain, even to the soft white chalk of the south-east, the oolitic limestone familiar to many as Bath Stone and the yellow scarps of the Magnesium Limestone of Durham. It differs from these in its hardness. It is a brittle and not a workable stone, as many other limestones, such as Bath, Portland and Caen stones. At Keynsham Abbey, where I have excavated for a number of years, the squared facing stones of ashlar are of Bath Stone, while the internal core of the walls consists of irregular pieces of Carboniferous Limestone and other rock. In spite of these differences, all limestones have one feature in common in that they can be reduced to lime by burning, and old limekilns are a feature of limestone areas. The vast quarries which are to be found in Carboniferous Limestone areas have caused quite a lot of anxiety to countryside conservationists who claim that they can permanently disfigure and alter the landscape. Although much of the Carboniferous Limestone quarried in the past was used for agricultural lime or for local building purposes, the construction of the motorway network brought considerable demand for road material and even some old quarries had to be re-opened to meet this demand.

Just as water action on the rock has left fossils in relief on cave walls, so are mineral veins often more visible in caves than on the surface. The constantly damp cave conditions darken the limestone, showing up the white calcite bands, and direct lamp lighting reflects the glitter of metalliferous particles in the rock which would not be noticed in the open air but is so obvious against the darkness of the cave.

The most familiar formations in caves are the stalagmites and stalactites, which, like the caves themselves, owe their origin to water action. The stalactites are those formations which hang from the cave ceilings and the stalagmites are the corresponding formations which grow upwards from the rock floors. Not all caves have these formations. Some are completely devoid of them, other caves have but a few, while in some the rock is barely visible through the richness of the stalagmitic formations. Parts of the same cave vary in this way. At Treak Cliff Cave at Castleton, Derbyshire, there is a fine display of these formations in one part of the cave only and even then, on only one side of the chamber, where the quantity makes up for the dearth of them elsewhere in the cave.

Stalactites develop as the result of drops of water on the

cave ceiling and as this water percolates through the joints and cracks in the roof, they usually consist of a string of formations along the cracks. As joints run usually parallel to each other, it is quite common to see stalactites in parallel lines across the ceiling. Stalagmites are formed from the residue of drips of water from the ceiling, so that many stalactites have corresponding stalagmites. In course of time, these may meet and join to make stalagmitic columns. Sometimes they form a 'grille' across a passage, such as the one to be found in the Grotto at Wookey Hole caves in the non-public part of the gallery.

The beauty of the underground scenery is often enhanced by wall formations, where stalagmite, the general term for the material which produces all formations, seems to have flowed down the walls and over boulders. The American call this 'flowstone', a term so expressive that it has been generally adopted by cavers.

A cave well covered with stalagmite is usually safer to explore than one in which the rocks are exposed, because the stalagmite seals the rocks together and a harder type of cement would be difficult to find. However it is not easy to climb, particularly when wet and slippery, for the hand- and foot-holds are rounded and shallow. Individual formations such as stalagmitic columns and particularly stalactites are brittle because of their comparatively small girth and cavers are expected to take particular care when near these formations; it is little wonder that many show-cave owners are wary of cave explorers. In a reputable club, it is a great stain on a member's reputation to break a stalactite, even by accident, for accidents are often due to mere carelessness, and in show caves and in wild caves the same care is exercised. Although 'a thousand years to an inch' rule is no longer accepted as constant, even a small stalactite may have taken thousands of years to form and can be destroyed in a second's carelessness.

Specimens of stalagmites and stalactites are to be seen in some museums and in show-cave shops. These have been found broken or recovered from a cave about to be quarried away or from a passage widened in a show cave. You have only to compare these specimens with stalagmitic formations in the natural state to realize the difference. It is almost as if a stalagmite were a living thing. Once removed from the cave, it seems to lose its 'life'. In the damp cave, it is glistening and full of beauty, but outside, in the open air and daylight, it becomes dull and stone-like. In a newly discovered cave, the purity of the formations is striking and the water in the cave pools may be so clear as to be invisible. If the caver is not particularly careful, the pools may become muddy and the formations tarnished. Certain passages found by Mr H.E. Balch and his colleagues at Swildon's Hole, Mendip, at the beginning of the century were said to be indescribable for their beauty, but even in the lifetime of the original explorers, much damage was done by later explorers and the passages could not be compared to what they were when first seen. While safeguarding the major formations, the caver must watch where he walks. The edge of a pool broken down by a boot may cause an escape of water and not only destroy the pool's appearance, but may be the end of a certain type of cave life in the pool.

Much of the attraction of stalagmite lies in the variation of its colour, due to impurities in the water from which it was formed. In its pure state, stalagmite is white, but this is rare. The general colour is a reddish-brown and in this case the water has percolated through iron-stained rocks or soils such as the Old Red Sandstone which often borders the Carboniferous Limestone. Green colouring and black are associated with copper and black manganese, but these colours can be due to a number of contaminating sources. In some caves the black is due to the sooty deposits from the days when cave guides would throw spirit over the walls and set light to it for illumination.

Water is merely the vehicle by which formations are brought about. They are, of course, of mineral origin, derived from the cave rock which they decorate, generally calcium carbonate of which Carboniferous Limestone is formed. The calcium carbonate found in cave formations is usually of the form known as calcite. As water penetrates down through the rock it takes certain of the calcite in solution and when it enters the cave, some of the calcite is deposited round the suspended drop. When the drop falls to the ground, the remaining calcite in the water collects on the floor. As each drop occurs, so additional calcite is added to that already deposited and so stalactite and stalagmite steadily grow. The spreading effect of stalagmites compared with stalactites is now apparent. Similarly, water flowing down a wall or over rocks leave wider deposits of calcite to form the flowstone.

Under normal conditions, the growth of stalagmite must be extremely slow and it will form only under certain conditions. It will not form in a situation which is too dry or too wet. Climatic conditions are therefore a controlling factor in their formation and, as climatic conditions do not remain constant, the growth of stalagmite is not continuous. There may have been very long periods when growth was in abeyance. The degree of solubility of the water, mineral content of the rock, the distance the water has to travel, the amount of carbon dioxide available from vegetation on the surface, all these factors will control up to mineral saturation point the amount of calcite in solution. The stalagmitic growth depends, therefore, not only on favourable climatic conditions but on other factors which vary from cave to cave. It is clear that we must accept with the proverbial grain of salt the usual statement made by guides in show caves throughout Britain that it takes a thousand (two thousand has been mentioned!) years for an inch of a stalactite to form. This theory is based on a deduction, made in good faith, in Kent's Cavern, Torquay. Here in 1688 a man named Robert Hedges carved his name and date in the surface of a stalagmitic deposit. By calculating the thickness deposited over the lettering since 1688, it was calculated that it would take a thousand years for an inch of stalagmite to form. The calculation was, of course, based on the assumption that the growth was a constant and continuous process, which we now know to be far from common.

During my archaeological excavation of Ogof yr Esgyrn (the Cave of the Bones) in the Swansea Valley, it was necessary to remove part of the stalagmitic capping of a mass of boulders. Between the boulders and on their surface beneath the stalagmite were pieces of pottery of Roman times, less than 2,000 years old, although the stalagmite overlaying them was four inches thick and, again, no doubt formed during only part of the time since the Roman occupation.

The ability of water to act on limestone depends upon its carbon dioxide content, a variable factor over the ages. During one period, decaying vegetation in the thick ground cover may produce considerable carbon dioxide, but plants are far more susceptible to even slight temperature changes than animals and in another period, different climatic conditions may mean less vegetation and less carbon dioxide. Then again, there is a periodic fluctuation in water action, for the more carbon dioxide it contains, the greater is the corrosive action and the more calcium carbonate it absorbs, but as the water absorbs more calcium carbonate, its corrosive action becomes weaker.

As we have seen, stalactites form from water drops on the cave ceilings, flowstone from the spreading of water over the surface of rocks and walls, and stalagmites rise from the cave floors. As well as these types of formations, practically every cave has stalagmitic curtains, thin draperies attached to the ceilings and walls, often looking like draped curtains at the side of a 'fairy grotto' tableau. They are due to water trickling along ridges on the cave ceilings. A curtain will often start as a stalagmitic fringe formed on the lip of a roof fissure. The fringe increases in depth as the growth develops. The deposits from the water on the ridge and from water trickling down the wall combine to form a small spandrel between the ceiling and the top of the wall. It increases in size due to water running down its edge to reach the floor. The irregularities during growth produce the familiar wavy edge. As water will have a tendency to drop where its progress is barred, a stalactite may develop on the lower edge of the curtain. Some curtains develop at the edge of large holes in the ceiling where water will tend to hang. An example of this is the 18-foot curtain high in the roof of the Cauldron Chamber at Dan yr Ogof caves.

It was once believed that stalagmitic deposits were formed by the precipitation of calcite due to the evaporation of water. However the high humidity of stalagmitic caves, particularly active ones, gives little opportunity for evaporation. Vapour is not usually seen arising from the rocks of a cave, except near the heat of powerful cave lighting or from the wet clothes of cavers when these clothes are being dried out by the heat of their bodies. It is now generally believed that the precipitation of calcite or other minerals—and there are some not so common as calcite—is due to the loss of carbon dioxide from the water to the air in a chamber. This levelling up between the water and air levels of carbon dioxide results in the deposit of some of the mineral contents of the water. This is because the saturation point controls the carbon dioxide level and any decrease means that the surplus mineral is deposited.

As the rate of deposit is variable, the deposits occur as layers which can be seen as bands in the broken edge of flowstone or concentric rings in a broken stalactite. The layers vary in colour and clarity, according to the minerals and mud in the water. As the water becomes clear of mud, so do the deposits. An example is the translucent 'bacon' appearance of the stalagmitic curtains.

The laminated deposits found over boulders and cave floors are exceptionally tough. Sometimes they have to be removed in the commercialization of a cave to increase the headroom for visitors by lowering the floor. These cuttings can often be seen in commercialized caves along passages where the exposed banded stalagmite edge can be seen along the walls.

On occasions, floor stalagmite has to be removed to assist archaeological or other investigation. At Minchin Hole, Gower, South Wales, a cave which I dug for some thirteen years, a very thick band of stalagmite can be seen projecting from the walls, high above the present excavation level. It is a relic of the excavations of the last century, when local squires vied with one another in the search for 'curiosities'. Its broken face is no longer fresh and its edge has lost its sharpness by the subsequent deposit of stalagmite, although today the cave seems exceptionally dry.

A small portion of stalagmite floor had to be removed on another thirteen-year excavation of mine at the Cave of the Bones (Ogof yr Esgyrn), Swansea Valley, in order to retrieve some Roman remains. Sledgehammers did not make any impression—they merely bounced off. Finally we tried 'feathering' with lump hammers and cold chisels. This is a method of turning the chisel as you strike it. The chisel just went on grinding a hole and not cracking the floor as we had hoped. At intervals the hole had to be cleared of the stalagmite crystals which looked rather like granulated sugar. It would have been easier if we had had an edge to work on, but the floor was continuous. Eventually the cold chisel fell through into a void between the flowstone-covered boulders. We put a crowbar in the hole and tried to lever the surface apart, but before we could be successful, we had to make two or three borings by feathering over a very close area. It is this almost impenetrable nature of a stalagmitic floor, particularly with primitive tools, that makes it such an important factor in the study of stratigraphy when dealing with prehistoric remains.

The variable conditions give some individuality to caves. Although the basic pattern may be the same, every cave has something different. At Castleton in Derbyshire, there are four 'subterranean wonders' in a string: Peak Cavern, Treak Cliff Cavern, Speedwell Cavern and Blue John Caves. Yet, if you visit every one of them, you will find some interesting variations in cave structure. Admittedly, lead-mining operations have added extra chambers to a number of Derbyshire caves or altered the form of some of the existing systems, but even in 'untouched' caves, known only to cave explorers, the variation is considerable. Swildon's Hole and Eastwater Cave on Mendip, Somerset, are within a stone's throw of each other, yet they are very different and Swildon's richness in formations contrasts with the bare rock of Eastwater.

As well as the major types of formations, the stalactites, stalagmites and curtains, there are still others, such as the 'fonts' at Cheddar Caves—a series of encrusted basins, rising in tiers and containing clear pools of water. These are formed by crystallization accumulating round the edges of rock pools. Such pools are known as 'gours' or 'rimstone' pools. 'Gours ' is a term borrowed from the French. Rimstone pools may be deep and consist of a series of deep steps, like the 'fonts' at Cheddar, or they may consist of shallow pools often in a string on a gently sloping cave floor. In many caves, these may be dry and their remaining evidence are the low banks of stalagmite which once formed their rims. As the floors of many show caves have been made up or bridged over, the shallow rimstone pools are probably a feature more familiar to the caver than the tourist.

Sometimes crystallization takes place on the surface of the rock pool and forms round, floating specks of calcite. Sometimes it will develop outwards from the side of the pool to form a thin ice-like sheet on the surface. As the crystallization increases, the small floating specks or the surface sheet of calcite will become too heavy for the water to support and will sink to the bottom when encrustation will continue.

Tiny rock fragments in shallow pools may have become rounded by the abrasive action of moving water. They attract calcite deposit and, being detached from the bottom of the pool, they look like pearls and are called 'cave pearls'.

'Straw' stalactites or 'straws' are long, thin, straw-like stalactites and as they do in fact consist of thin tubes, the term is appropriate. In certain parts of Dan yr Ogof caves they appear as a veritable forest on the ceiling. During the re-exploration of the caves in 1937, we had to be exceptionally careful when passing through areas not to accidentally brush against the thin white 'straws'. Fortunately, there is a fine cluster in the public sector of the caves, well up in the ceiling and out of harm's way, but visible for all to see and known as the 'pin-cushion'. It is from the straw stalactites that the normal stalactites develop. Why should the straw be tubular and the fully formed stalactite solid? For the answer we must go back to that first drop on the ceiling. Because of the escape of carbon dioxide, a skin of calcite forms on the surface of the drop. As the drop increases in size, the calcite skin bursts and allows the water to fall to the floor. A small rim of calcite remains round the neck of the drop and, as more water percolates through the orifice, more calcite collects round the edge of the rim, which gradually develops into a thin straw. With a reduction of water, crystallization may occur within the interior of the tube and water is forced to flow down the outside to form a mature solid stalactite, forming rings of deposit.

Sometimes, old breaks are found in formations, even in

newly discovered caves. These are often attributed to old earth movements and a new stalactite of small dimensions grows on the stump of an older one and sometimes a new stalagmite on an old stalagmite stump.

Some stalactites are unusual in being twisted or distorted. They are known as 'helictites' or 'eccentrics', as they do not stick to the rules. Various theories have been put forward to account for these eccentrics, including draughts not strong enough to remove the drop of water, but enough to keep it to one side. The only problem is that an eccentric may be found as one of a group of normal formations and do not themselves appear to be common in groups. Experiments have shown that in some instances, a ruptured wall of a blocked straw will cause side excrescences. Sometimes the shapes of helictites are so fantastic that they acquire names like 'the little man' or 'the pig's tail'. In some cases, they may be due to the formation of calcite on roots or even fungus.

Although calcite is the normal deposit, other minerals are also to be found as crystalline deposits in caves. A general term for all precipitated deposits, including calcite, is sinter. It includes many forms of deposit, including that which collects round a bowler hat, or whatever you may choose to place in those surface streams reputed to have a 'petrifying' effect on anything placed in them. 'Petrification' in those instances is, of course, not a correct term, as the object itself is merely coated.

What sinter deposits other than calcite are to be found in caves? The one other form of calcium carbonate in caves is aragonite which in appearance cannot be distinguished from calcite, but has a different chemical reaction in the laboratory. Gypsum is another substance which is reasonably common, sometimes in formations, and various other minerals are to be found in small quantities.

THREE

Exploring Caves—Past and Present

IN THE EARLY PART of the century, when cave exploring was a new sport, the equipment considered necessary was little. There were no wet suits and no helmets. Any old clothes were worn and the headgear was often a battered hat. During my early caving days, I wore an old turned-down trilby, pulled well down to stop it coming off, and later added a small acetylene lamp, of the type once used on bicycles, by tying it to the hat band. I also wore an old shirt, a pair of plus fours, heavy socks and walking boots. For extra warmth, I wore a woollen V-necked pullover and woollen underwear. The old type of long pants were an asset in those days, but the main thing was old clothes, anything too old for ordinary use. The plus fours were an advantage over normal trousers which were inclined to get heavy with water especially round the turn-ups. They were also an advantage when climbing a rope ladder as they were not so likely to get caught in the rungs and gave you a clearer view when trying to get your feet on an awkward rung. Our attire was completed by the addition of a short piece of coiled rope round the waist or diagonally across one shoulder. It looked professional and you were never sure when you might need it in getting down short inclines.

As the supply of old clothes wore out, we progressed to one-piece boiler suits and these became the normal caving gear. They were less cumbersome and there were no pockets or loose jacket to get hooked up in a tight passage. The change to boiler suits was quickly followed by the miners' helmets. This was the first item of safety clothing to be introduced into the caving world. Through various club journals, all cavers were urged to acquire one and they soon became essential, so there was a spate of black helmets with small electric lamps either on an elastic hat band or in a fitting on the front of the helmet. Many cavers added chin straps to prevent helmets coming off in narrow passages or falling down drops. I had not owned mine for more than a week when a small rock fell on it as I was making the ladder descent of the newly found Cow Hole on Mendip, a small rock but large enough to make a permanent dent in the helmet instead of my skull.

Although we had these small and not always reliable hat and helmet torches, candles were still the main source of illumination when I first caved. Everyone bought bundles of number eights or twelves, that is eight or twelve to the pound weight. Mine were mostly twelves because they were thinner and took up less space in a small bag with sandwiches, hand torch and spare batteries. The candle gave an immediate good all-round illumination, but was limited in its range. The light was mellow compared with the whiter torch which was used for looking down holes and into passages, as the candle was useless for that purpose. One of the disadvantages was that candle grease would be left on rock surfaces and added to the mud or your clothes. It would drip on your hands, but the hot candle grease and warmth of the flame would be quite welcome when you were wet and cold from the cave stream. The candles grew hotter to hold as they burnt down and were always renewed before attempting a long tight passage as it would be difficult to get out a new one from the supply in a narrow space. The new candle was lit from the flame of the old one and stuck on the melting stump. It was not wise to carry non-safety matches as they might ignite from friction as you dragged through a low space. More than one would-be explorer was rescued from a cave without lights because of wet matches, and many carried them in a tin sealed with surgical tape. It was often said that torches might fail and often did, but as long as you had candles and dry matches you always had a light.

Sometimes the candle did fail, particularly in a draught, and if this happened in the drain-pipe in Goatchurch Cave, Mendip, there was little chance of relighting it. The 35-foot crawl was always difficult with primitive lighting. There was just sufficient room in this round drain-like passage to lie flat with arms stretched out and push along with your toes. You couldn't turn round and come out until you reached the small terminal chamber. There was a shallow dog-leg bend in this passage and it was there that you always seemed to meet the draught. If your candle went out there, there was no room to get at your matches and you had to rely on the glimmer from the person in front or behind.

The essential equipment for descending cave pitches is a pliable ladder which can be rolled up, carried through the cave passages and used to descend vertical drops within the cave as well as deep entrance shafts or potholes. Today these are made of wire cables with metal rungs, but the early ladders were ropes with heavy wooden rungs.

My club had no ladders when we first went caving, but three of our members owned a building firm and they presented us with a supply of ropes and a four-fold pulley block, plus a quantity of rope bought from a ship's chandler. We used the ropes for climbing and when we were faced with a dry pitch, we used a bosun's chair—a short plank as a seat, with a hole at each end, through which the rope was threaded and tied in a loop. The end of the rope passed over a single pulley wheel tied to scaffold poles borrowed from the builder's workshop. Occasionally we would dispense with the pulley and the bosun's chair was lowered or raised by hand using a strong team of cavers. Another team controlled the lifeline, a second rope which was tied under your armpits with a bowline knot. We had to know our knots fairly well and a feature on knots was a common subject in the journal we issued to members.

The advantage of a pulley and ample rope was that the bosun's chair and lifeline could also be operated from the bottom of a pitch. A whole party could make the descent, leaving one man at the top to make sure that some irresponsible person did not come along and interfere with our tackle. A problem of having two separate teams for the bosun's chair and the lifeline was co-ordination. At Coral Cave, Wavering Down, Somerset, when I was being lowered on the bosun's chair, the lifeline was not paid out at the same speed and I was lifted out of the chair just at a point when the chair got caught against the wall of the cave. I swung on the lifeline in an arc and had to twist round to ward off the oncoming wall at the other side of the chamber with my feet. The impact set me swinging back the other way and as I passed the chair, I tried to grab it, but all I did was to dislodge it and it swung in the opposite direction. The bosun's chair and I were two pendulums swinging in our own arcs. I tried to catch it on the return journey but missed. On the next encounter, I managed to catch it and, after a lot of shouting, I finally managed to get the lifeline slackened so that I could get into the chair and finish the descent. In fact, it would have been far easier to have been lowered on the lifeline, but we had a firm rule that there must be a second means of support.

For climbing purposes, we had two bell ropes complete with their furry 'donkey tails' in red and white, although they soon became stained with mud. They were nice cotton ropes and of good diameter for climbing. They were too big for narrow cave passages when rolled up and usually had to be dragged through the mud and water and were known as the Loch Ness monsters.

Like most cave clubs we made our first rope ladders. We discovered an old seaman who knew how to make them and he spent many an hour teaching us how to assemble our equipment. We had a number of hickory rungs, cut and notched round both ends. Then dividing the three-strand manilla ropes, alternately one strand one side and two strands the other, we inserted the rungs and whipped the ropes with marline above and below each rung. Then the 'tails' were spliced round metal thimbles from the ships' chandler to be linked when in position by shackles. He was a good teacher and by the time the ladders were finished, he had taught us a wealth of information about knots, ropes and splicing, so that armed with a marline spike, we were able to carry out our future repairs and make further ladders. I think they were made for a lifetime because one of them reappeared in a museum exhibition tableau in recent years, complete with a model of a caver climbing in a slouch hat and old clothes to show what the early days of caving were like.

The ropes were treated with Stockholm tar, a pleasant healthy smell in the confines of a cave. Unlike modern metal pliable ladders, they would stretch and creak when you first stepped on them, a rather hair-raising experience for a beginner when making his first descent. As he stepped on the top rung, the tether ropes, usually tied round a handy stalagmite or a bar driven into the rocks, would give and stretch as would the ladder sides so that the rung would go down under his weight.

There were two disadvantages. The ladders were wide, making cumbersome bundles, and caving is exhausting enough without the burden of heavy equipment. They were made to take two feet on each rung. In a narrow passage, the old ladders would sometimes obstruct the way ahead and have to be pushed along or dragged behind. I have known more than one occasion, in a narrow rock channel, when a pile of ladders, usually passed through piece by piece, accumulated to such an extent that the passage was completely blocked and there was practically no space in which to handle them.

The next style of ladder was narrower and took one foot only, a precaution in case the wooden rung should break, and so was easier to carry. The ladder had thick flat rungs, bored with holes at each end, through which the ropes were threaded and secured by copper pins through the thickness of the rungs and passed between the strands of the rope. The ropes were then whipped with marline, just above and below the rungs. There was an advantage in the ropes being threaded through holes in the rungs in that, although the rungs might eventually show signs of wear through contact with the rocks, the ropes were protected from chafing from the same cause.

The ladder sections, perhaps fifteen- or twenty-foot lengths, were linked by karabiners, a kind of spring hook, in the form of oval links. Those first used were of Swedish steel and were an excellent safety factor. They were also useful worn on a caver's belt or waist rope, because all he had to do to rest at any time was to press it against the rung for it to engage and hold him on the ladder. Like some other cave techniques, it was borrowed from the climbing world to which many cavers belong. For a time, karabiners were looked upon with some suspicion, as a spate of cheap karabiners came on to the market with a low tension factor, but good karabiners were, and are, a valuable addition to the caver's equipment.

The rope ladders were superseded by pliable metal ladders, usually consisting of light metal alloy rungs, with galvanized steel flexible wire sides and the tails fitted with thimbles. The technique of using this ladder was borrowed from French caving circles and was far less bulky and heavy than the rope ladders. It is the type of ladder used today. They were termed 'electron' ladders as the rungs were of electro-magnesium aluminium alloy and, although other metals or alloys are now in use, the term 'electron' has come to be used for any similar lightweight ladder.

Lighting equipment has also improved. The candles gave way firstly to the acetylene head lamp, which gave a mellow all-round light, but the burners often blocked. Water was not always handy in dry caves, and worst of all, you had to get rid of the spent carbide. Many indiscriminate cavers would leave heaps of foul-smelling residue behind. They were also inclined to blacken the roofs and walls. Large acetylene lamps are still used by the French and Spanish prehistorians when examining and searching for prehistoric engravings in caves, as these lamps give better illumination for the purpose than electric lamps, but of course, the prehistorians are quite aware of the irreparable damage they can cause and know how to use them with caution.

When caving became popular, the old type acetylene lamp once used on bicycles came into their own again for a time, but when they fell out of favour, they found another market with target pistol clubs as an excellent means of temporarily blackening worn gunsights. The soot from the flame gives a flat matt finish which does not reflect light. The electric lamp also has improved and has superseded all other forms of lighting. Today they are generally of the type used by miners and are capable of being recharged. The usual type of hand torch, unless of a thoroughly waterproof type, has a number of drawbacks under damp conditions, and the caver, of course, needs his hands free.

With the continual experimenting with cave techniques, the tendency has been to smoothline the caver's progress as much as possible—no more bulging pockets to get hooked up in tight places, no flaps or belts to get caught on projections, or trouser legs to hamper movement and, above all, no longer the cold ordeal of struggling through water and emerging to climb and crawl with oozing dripping garments.

In spite of bulky tackle and primitive lighting, the greatest caving hazard in the old days was cold water. We piled on the woollens and these helped because when they became soaked, they held the water next to the skin and the body warmed it to a certain extent. Today neoprene wet suits retain the body heat. They are streamlined and comfortable, too warm at times, and a far cry from the old clammy, dragging garments. One result is that many members of a caving team look alike in their smooth, black and even tidy suits. A recent development has been the waterproof boiler suit which some cavers prefer.

These suits have greatly reduced the danger in a wet cave of collapse and even death from prolonged exposure (hypothermia is the medical term). The fairly even temperature of caves, varying but little between summer and winter, makes caving a round-the-year sport. It is a relief in summer to escape the heat by entering the cool depths of a cave, and in the cold winter you may feel the warmer air from the cave entrance. However a soaking in a wet cave quickly reduces the body's heat by evaporation and unless the heat is replaced by exercise or the loss of heat prevented by insulation, exposure can result. Occasionally a caver may be unable to move through an injury caused by a fall, or on a too-long caving trip he may become exhausted. The heat generated declines and what has been stored is lost through evaporation. Shivering and confusion occurs, followed by semi-consciousness, when the shivering ceases and is replaced by stiffness of the muscles. If the heat loss continues, the next stages may be unconsciousness and then death. Some clubs and cave owners insist that cavers wear wet suits when exploring river caves.

A considerable number of techniques are used in the course of cave exploration, as a cave system can pose quite a number of problems, each having to be tackled in a different way. In fact, the challenge of caving is that in an unknown cave you never know what is round the corner and what kind of obstacles you may have to overcome. The climbing techniques are similar to those used in mountain climbing

Cheddar Gorge

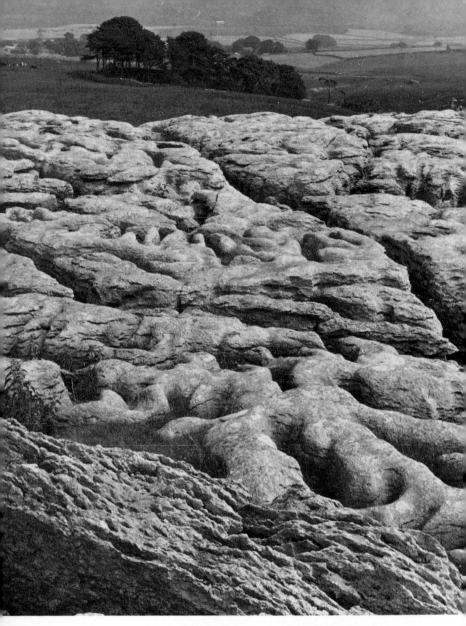

Limestone formation, Borrin's Moor, Yorkshire,
with Alum Pot in the distance

Castleton from Mam Tor

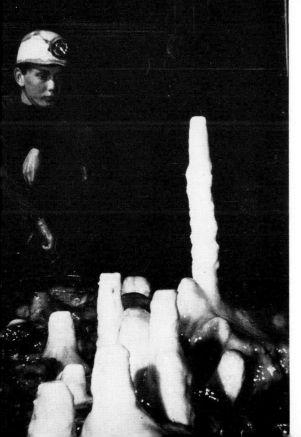

Stalagmites in Straw Chamber, Dan yr Ogof

Curtain formation,
The Cauldron, Dan
yr Ogof

Straw stalactites,
The Canyon,
Dan yr Ogof

Early caving days—
the author in
Swildon's Hole,
1934

Climbing in Cathedral
Cave, Dan yr Ogof

Ladder descent into
the Abyss, Dan yr
Ogof

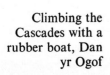

Climbing the
Cascades with a
rubber boat, Dan
yr Ogof

Greater Horseshoe bats in a Devon cave – a
species protected by law

Cave surveying in Dan yr Ogof

Elephant's tooth in
Joint-Mitnor Cave,
Buckfastleigh, Devon

Bone stacking by
Beard, Banwell,
Mendip

and in fact, caving has been described as mountaineering upside down. Many cavers pursue both sports and some caving clubs have a climbing section. The terms used and basic equipment are similar. As in mountain climbing the amount of equipment which can be carried is limited and space restriction is a factor which comes into both sports. Obviously there are considerable differences. The mountaineer does not have to crawl through underground torrents or run the risk of being trapped by floods and his sport is carried on in daylight. On the other hand the caver is protected from the weather and temperature changes in the cave and although he has to watch rise in water levels due to torrential rainfall outside, he does not have to worry about gale-force winds or failing daylight. The rewards, too, are some similar and some different. They share the same zest of physical effort and the excitement of a new environment but the climber gains his view of the vast panorama of the mountains, while the caver has the limited weird world of the caves.

As in climbing the most essential items in caving equipment are good ropes. Most of the ropes during my early caving days were of manilla with an occasional cotton rope, but the ropes today are often of nylon, which are not so affected by the alternate wetting and drying processes which can be so destructive to manilla rope. With the more general use of artificial fibre, in addition to nylon, plaited polypropylene and terylene are also used in rope-making. One of the greatest dangers apart from strength is the risk of abrasion to a rope, and this, perhaps the principal cause of breakage and wear from rock contact, is not always easy to detect in artificial fibre ropes. All ropes vary a great deal in quality and there is still a lot to learn about the qualities and faults of man-made fibres and their reaction chemically and under stress, when subject to certain conditions. Research is still progressing, but whatever material is chosen, only the most suitable and to a standard required for cave exploration should be purchased. A test carried out on a polypropylene rope showed that abrasion was assisted by the low melting point of the fibres under friction and although nylon also suffered in this way, the effect was much slower. The method of construction of rope has some bearing on the strain on man-made fibre ropes, particularly when subjected to sudden excessive pulls. Care must be taken, when using man-made fibre ropes for climbing and particularly for abseiling, as these ropes lack the stretching properties of natural fibre ropes, which take up some of the shock in a sudden jerk, with the result that if not sufficiently robust they could snap. Man-made fibres, too, some more than others, are susceptible to chemical damage by contact with what might appear to be the most innocent of materials, even sunlight.

Since ropes are such an important part of a caver's equipment, it is essential that they are not only treated with care when in use, but their maintenance and storage when not in use is equally important, as ropes, including those of artificial materials, can deteriorate. Most cave clubs have an equipment officer, whose job it is to maintain the ropes and other equipment and to report to the club committee the state of the tackle at frequent intervals. He loans out the tackle to caving teams and checks the equipment on return. Some cavers prefer to use their own personal ropes, but this depends on the rules of the club, who, of course, may have some responsibility where official club trips are concerned. Most clubs carry insurance and have to comply with various conditions laid down by insurers. One of the problems of ropes and rope ladders used to be the possibility of damage by nailed climbing boots, but this has been reduced by the general use of composition climbing soles.

Caves are rigged, that is, the tackle placed in position for each cave trip, and although equipment has been left in caves for the next expedition, it is not good practice, for apart from unnecessary exposure and increased deterioration, its presence can encourage amateurs to make descents for which they are neither sufficiently experienced nor equipped. There is a risk that another party might enter the cave while exploration is in progress and descend a pitch and be in part of a labyrinth below the drop, when the original party makes the ascent and removes the equipment. It is always wise to leave a member at the top of the pitch to make sure others do not use it or interfere with the belays. Such a risk is not so likely today, as many caves are gated and the keys held only by competent clubs, who are aware of such possibilities.

In some cases equipment has been permanently installed in caves, where it is thought to be of advantage, such as the rigid metal ladder in the entrance shaft of Lamb Leer, Somerset, and certain cables and chains on some awkward sectors of Ogof Ffynnon Ddu, Wales.

Every caver must know the technique of climbing a pliable ladder, which is only secured at the top to a belay either natural, such as a stalagmite or rock, or by a metal belay, often a bar driven deeply into a crevice in the floor. The ladder hangs free and has a tendency to swing and twist as soon as you put your weight on it, except when in direct contact with the rock. A free-hanging ladder is not easy to climb for the first time. Your feet push the ladder away from under you and as you hang on with both hands you suddenly see the toes of your boots come up almost to the level of your face and further progress becomes extremely difficult, to say nothing of twisting. The technique is to keep the ladder as rigid as possible and this can be done by descending down one edge of the ladder, holding the far edge with both hands and placing the toes on each successive rung from opposite sides. An alternative method is to descend on the face of the ladder by using the toe on one rung and the heel inserted in the next rung from the back of the ladder, but this is difficult and slow. In all cases a lifeline should be used paid out by a team at the top or controlled from the bottom by passing over a pulley suspended from the top.

There are some places in caves where the caver wishes to get up to a higher level and where because of the narrow approach passages, it is impossible to use a rigid ladder. It may be possible to climb the rock face using pitons if necessary—metal spikes driven into the rock by the climber—and then hoist the ladder up by a line and belay it. In the case of an unclimbable face, the explorers must depend on other means such as a 'maypole' which consists of a number of metal tubes which can be clamped together on site with a crossbar on top, to which a pliable ladder is hung. It is then erected so that the top is at the level to be explored and the foot at an angle to the face, placed against boulders or the opposite wall of the chamber. Guy lines attached to the ends of the crossbar are fixed at suitable angles from the base to stop any lateral movements of the pole. The ladder hangs down the wall.

As the drawing up of a ladder from the top may cause unnecessary abrasion when it is pulled over the ledge, the bottom may be drawn up first by a rope from the top. This sometimes prevents the rungs being caught on projecting rocks which often happens when a ladder is drawn up direct. Sometimes a face is so irregular that the ladder has to be unhitched by a man, suitably secured by a lifeline on the ladder itself.

Metal tubular poles and clamps, similar to those used in scaffolding, are a standard part of the equipment of a number of cave clubs, for they are useful for bridging gaps, erecting safety barriers, for belaying purposes and for such things as winch platforms. One of the most notable uses of a winch is at Gaping Gill, the famous Yorkshire pothole. Winches are used to raise and lower platforms, buckets and bosun's chairs, where it is necessary to lower equipment, bring up debris or where the depth is so great that the descent by ladder would be tedious and time-consuming.

The number of techniques used in cave exploration is always increasing. In the 1970s, we find abseiling taking the place of ladders. This method, used considerably by mountain climbers for many years, became popular among a number of cavers, perhaps because cavers and mountain climbers seem to have become more intermixed. Abseiling is a descent by a single rope which is passed round the body in a certain way to check the rate of the descent. Today, certain harness equipment can be bought which checks the rope mechanically and in some instances enables the user to ascend in the same way, using the feet to 'walk' on the vertical rock face. It was during an abseil that David Huxtable fell to his death in Gaping Gill in December 1974, not due to a mistake in the technique, but due to the rope breaking when he was 50 feet down the 350-foot descent. At the time of writing it was the first and only fatal accident in a cave with an abseil.

'Abseiling' is normally used in describing the form of descent on a single rope, but as the method is now also used for ascents, the term 'single rope technique' is used to cover both. The descent is still termed abseiling, but the ascent is called 'prusiking' after its use by the Austrian mountain climber, Dr Prusik, for rescues from crevasses. Although the system first became popular in the first half of 1970, it had been in use by mountaineers and certain cavers for some years. In its simplest form, it consists of a number of slings fixed to the rope by special knots which tighten when weight is put on the slings, but free to slide when the weight is reduced. A more advanced method is the use of a specially made harness which carries out the same operation and is often known as a 'monkey'. A simpler form for descent is known as a 'descendeur'. Quite a number of varieties of commercial equipment for abseiling and prusiking are available for purchase.

Water is a problem in many caves and varies from region to region according to the proximity of the systems to the water table. Wales is a land of particularly wet caves, 'living systems' which still contain active streams. Where the water is reasonably calm, as on certain subterranean lakes, canoes or, where access is restricted, inflatable rubber dinghies can be used. At Dan yr Ogof in Wales, a coracle was actually used during the early explorations at the beginning of the century. Later explorations of this cave saw the use of canoes and inflatable dinghies. Two amusing episodes can be told of their use. On one occasion, two cavers were paddling along in the tunnel connecting two of the lakes. Their canoe had been holed in the prow, but by sitting in the back, they kept the hole out of the water, for a time. Presently the backwash of water from the walls of the tunnel began to fill the boat. They still went on paddling, sinking lower and lower in the water until the boat disappeared below the surface. It then sunk from beneath them and they had to swim back to the tunnel entrance.

The second occasion concerned an inflatable dinghy. It was during the war years and the leader of the explorations at Dan yr Ogof at that time was Gerald Platten. He and several of his team, still helmeted, were travelling in a car some miles from Dan yr Ogof where they had been exploring when they were stopped by an auxiliary policeman. The helmets had attracted his attention, but the presence of two German-made cameras convinced him that they must be enemy invaders. When he went round to the back of the car and saw the rubber dinghy tied on to dry and still wet, he had little doubts that he had caught the enemy red-handed. They had quite a job talking themselves out of that situation.

Not all water in caves is calm enough for navigation, as for example in the stream passage at Ogof Ffynnon Ddu, where the water rushing down the slope can sweep you off your feet. In such cases, the caver has to fight against the force of water when conditions allow. In some instances, permanent chains or cables are fixed to the walls as hand holes as in Ogof Ffynnon Ddu. In some show caves platforms are sometimes built over streams, and in others the stream is kept to a defined channel, leaving part of the passage floor dry. Artificial tunnels have been made to avoid such areas, as in the show cave of Dan yr Ogof, Wales.

Some years ago, a member of a British cave team, visiting France, lost his foothold while they were trying to force their way up a river exit of a cave. The river swept him out of the cave and along the valley outside, a fairly rough journey as the river made its way between numerous boulders. The caver managed to get a hold just below a road bridge, where a number of people had seen him struggling in the river. They were amazed to see him crawl up on to a rock, water pouring from his clothes and his pipe still in his mouth. Needless to say, the French press made much of the incident and of the intrepid Englishman who, at the risk of drowning still held his pipe in his mouth. The caver later told his fellow cavers that the only reason he had held on to his pipe was that he had to grip on something.

In some caves, passages and even chambers, as at Wookey Hole, Somerset, can be completely waterlogged and sometimes, cave passages can be partly filled with water to such an extent that there is no room for the caver without submerging. The submerged distance may be quite short and involve only a quick dip from the air space on one side of a rock to the air space on the other and is usually known as a 'duck'. The submerged or almost submerged passage may be quite long, often dipping down into the water and rising again some distance ahead. All such passages were once known as syphons, but as they were not true syphons, they are called 'sumps'. In some cases they have been swum by holding one's breath, known as free diving, but there is a limit to this method and if the distance is too great, the caver has to obtain an auxiliary air supply by diving techniques.

For many years, further exploration in a number of caves was held up by the presence of a water barrier. The amount of potential exploration beyond these barriers was enormous and has proved to be so, due to the early efforts of Balcombe and Sheppard, who later founded the Cave Diving Group, which was for many years to embrace the whole of cave diving in Great Britain. It still holds pride of place, in spite of some individual and small group diving in the last few years, and has branches or groups in the major cave regions.

Although the Cave Diving Group was formed in 1946, Balcombe and Sheppard had been diving for over ten years. The first public appearance was at Wookey Hole Caves in 1935 after intensive training at Waldegrave Pool on Mendip, under an instructor of Siebe Gormans with whose help they were able to qualify as divers. The exploration at Wookey Hole Caves in 1935 was a large-scale affair. The suits were of the deep-sea diving variety, with copper helmets and heavy lead boots and were linked by telephone and air tube to a manually operated pump from the first dry chamber. The suits were clumsy for cave diving, as was the air pump operated by a bar by two teams of opposite helpers, rather like the first manual fire pumps. Because of the noise of pumping, the pumps had to be stopped when the telephone was in use. A description of the operation, with commentaries from the divers, was being broadcast on sound radio at the time. The diver's voice became fainter as the air in his suit decreased and gradually trailed off, ending with a distant voice asking for more air.

Apart from the clumsiness of the suits and air pumps, the air feed was a drag on progress and had to be lifted by the divers and eased round projections, becoming heavier to handle the further they were from base. However the equipment enabled them to see some of the green underwater world of Wookey Hole, which had been unknown at that

time. The next operation was through the narrow sump of
Swildon's Hole a few miles away from Wookey Hole, but this
time they had a home-made apparatus as the passages were
restricted.

The war years brought with them the lighter frogman type
of self-contained equipment fed by oxygen bottles, carried on
the person. The diver inhaled almost pure oxygen and the
carbon dioxide exhaled by the diver was absorbed by a
container of soda lime, carried on the harness to which the
oxygen cylinders were strapped.

The suits varied in design. On my own first training session
at Ffynnon Ddu, I was given a rubberized canvas suit. You
entered it by a rubber funnel which protruded from its
abdomen and, after wriggling your body into the lower and
upper half of the suit, your face emerged in the opening of
the rubber hood and your wrists through the tight rubber
cuffs. The funnel was folded up and sealed with a large screw
clamp and band of rubber.

The weight belt was fitted with a series of pockets, each
with a slab of lead. In those days, all exploration was done by
walking the river bed and the weights had to be adjusted for
each person. The aim was a negative buoyancy, just sufficient
to enable you to walk the river bed, without being too heavy
or too light. To keep you as vertical as possible, lead was
placed over the soles inside your boots. If you wished to
surface in an emergency, you jettisoned sufficient of the
weights to enable you to ascend and this was done by
releasing press buttons on the straps under the pockets.
Where there was water surface open to the air, this
arrangement was fine, but many underground chambers are
water filled and the likelihood was that you would float up to
the cave ceiling and remain there, while other divers searched
the river bed below you. Because of the possibilities of these
and other emergencies, it was and indeed is a rule of the Cave
Diving Group that nobody should dive alone.

It was in this kind of equipment that I stood up, ready for
my first dive, on the green soggy banks outside Ffynnon Ddu,
contemplating the dismal depths into which I was going to
descend. I was wondering whether I had done the right thing
in volunteering when a nose clip was put on and the oxygen

gag thrust into my mouth. It was too late to talk and I heard the hiss of oxygen when the bag filled up. I was told to do my 'breathing drill' to expel any air from my lungs. "Now breathe normally" came the request. This seemed difficult with a gag in the mouth and a clip on the nose. A large pair of goggles was put over my face, accompanied by the words, "Follow me."

The weight was terrific. Suddenly I felt the coldness of water round my legs as I followed the instructor into the cave and saw him disappear beneath the surface of the water. I did the same and was amazed how weight had suddenly disappeared. I was just beginning to feel at home in this new world, when I followed the instructor out of the pool. I had forgotten about the weights. My knees gave from under me and I rolled over on my back, as helpless as a medieval knight when he dismounted in full armour. Helpers seized me and the dismantling process began. When I had been finally skinned out of the suit, and stood up in the special woollen one-piece undergarment known as GNU, 'Gents' Natty Underwear', I felt lighter than I had ever been.

For some years, modified forms of this type of suit with oxygen breathing apparatus was used because oxygen cylinders would give the same breathing time in a much smaller container than the air cylinders, but difficulty arose in exploring caves with really deep water, since oxygen becomes toxic over a certain depth. The safety factor with air cylinders is greater too and they are now widely used, as are flippers instead of the heavy boots, so that much greater distances are covered.

The elements of danger which exists both in caving and in diving are combined in cave diving, so the Cave Diving Group has always been much concerned with safety and proper methods of control. They have always been careful in the selection of their members and they are drawn from experienced cavers who know about the problems of caving before they embark on training as divers. The Group is now divided into teams working in the various cave areas of Britain.

The accessible parts of practically all known caves have already been fully explored and the Cave Diving Group are penetrating water barriers into new passages and chambers.

It is the ambition of most keen cavers to find completely unexplored new systems and so they turn their attention to swallet digging. The existence of underground systems is usually apparent on the surface. Walking over the uplands of limestone country, one notices large hollows sometimes with a stream going underground. These cone-like hollows are called swallets, swallow holes or sink holes. There is often a string of them, some active secondary swallets and some dry depressions, where there has been some internal collapse of the cave. A series of swallets usually indicates a sizeable underground system and the course of the cave can be traced by following the line of the hollows.

A likely place to dig is where a stream goes underground at the bottom of a swallet. These stream entrances rarely remain open. Silt and stones are washed into them and the sides of the holes break away, eventually forming the funnel-shaped hollows. It might appear a reasonably easy matter to remove the boulder choke and enter the stream passage, but it is usually found that the boulders extend far into the system. It can be a difficult and hazardous task to make a way through them. Lifting tackle and shoring are often needed.

Swallet digging may involve years of labour and I suppose most people who have been caving for any length of time have worked on a swallet dig which did not 'go'. When the rare breakthrough does happen, however, the caver has the unique reward of exploring unknown territory, where nobody has ever been before—not on the other side of the world, but under some hill in Britain.

FOUR

Cave Studies

MOST PEOPLE go caving for sport in the same way that others go climbing, canoeing or surf-riding, but many sporting cavers sooner or later take an interest in some other aspect of caving, or speleology as the cave sciences are called. Some join a caving club which caters for such interests and begin to specialize in a particular branch of cave studies.

The study of wild life is always absorbing and many cavers are fascinated by the wild life which exists in caves. The most obvious creature is the bat. It is a mammal not entirely restricted to caves, but caves are favourite retreats for some of the species. To many people bats are just ugly, cumbersome and even loathsome creatures associated with vampires, graves and beasties of the night. It is a pity that popular fiction has cast bats in such a role for they are fascinating to study. We have, of course, no blood-suckers of animal herds in this country. British bats are timid, insect-eating, furry little animals which shun the light and strangers. The idea that they can get caught up in people's hair is another myth. There may be the very occasional incident when a bat can get caught, but it can manoeuvre in surprisingly small spaces, without touching anything. It is a better master of flight than any bird for its wings are relatively large to its body, enabling it to turn, twist and drop in far less space than a bird requires. Its very elongated fingers are connected by an expanse of wing membrane which terminates at or near its ankles, according to the species, giving a relatively enormous expanse of wing to produce great manoeuvrability which is essential to an animal which feeds on the wing. The apparently erratic flight is due to the ability to dart after flying insects in practically any direction. Often when seen in this way, it is feeding and provides evidence for this by rejecting insect wing cases and other unpalatable parts which fall to the ground.

The advantage for flying of wing membrane between the fingers and ankle and between ankle and ankle turns to a disadvantage when it comes to walking. In fact the animal is incapacitated as we would be if our ankles were shackled together. The bat is adapted entirely for flying. It lives either in flight or suspended by its tiny feet from a ceiling sufficiently high to enable it to drop and take to flight while in the air. If a bat is 'grounded' it will pull itself along by its wings in an ungainly fashion, until it can fall from a ledge when it will take flight during the fall. In fact, a bat struggling on the ground can easily be mistaken for an injured animal. When suspended, the bat folds its wings against its body, but the horseshoe bat wraps them right round, so that colonies of these bats look rather like masses of dark tulips growing upside down from the ceiling.

Bats are disturbed by light and if awakened in a cave by cavers' lamps will seek darker areas for retreat. In the open they are not seen in flight before dusk and this is no doubt the origin of their evil reputation in fiction. Although they have eyes, they operate by sound and can therefore manoeuvre rapidly in darkness. It is believed that bats pick up sound from walls and rock obstructions as bats have a built-in type of natural echo-sounding system. Sounds given out continuously by bats, outside the human hearing range, are reflected back from surfaces by echo—much on the same principle as the audible echo of man-made systems used on ships to locate submarines and obstructions.

The fact that bats depend upon sound and not sight for guidance was first discovered by Spallanzini in 1793. He discovered that blindfolded bats found their way about with no less difficulty than other bats, but when their ears were plugged, their sense of direction failed. It has also been found that the failure will occur if the ears are unplugged but the mouth and nose covered. The mouth and nose are the transmitters and the ears the receiver. It is now known that a bat sends out sounds that can vary from within the human ear range to the ultrasonic. The faster the bat flies, the more rapid are the number of emissions it makes per second and also the closer an obstacle, the more rapid they are, like the sounding apparatus on a ship.

The normal cry of a bat can be heard by human ears as a high pitched 'tweek', but the ultrasonic vibrations are produced through an opening in the larynx by a flow of air which increases with flight, but still operates at a lower pace when the animal is at rest. The pitch of the sound varies between species. These vibrations can be converted by electronic equipment into sounds which can be received by the human ear. This equipment is useful, not only for studying bat transmission but for locating bat colonies. An advantage of the electronic system is that it avoids disturbance to bats by handling, because a species can now be recognized in flight by the transmission it makes. It is not easy to recognize bat species in flight by normal observation and previously it was necessary to study the animal by handling it. This may have led to a decrease in species, which is discussed later in the chapter.

Until 1935, it was believed that only bats were guided by an ultrasonic system and were unique among mammals for that reason, but in that year American researchers revealed that dolphins also operated on a similar 'warning system'.

The largest of our cave bats is the Greater Horsehoe with a wingspan of 12 to 14 inches, which might seem quite large when compared with our smaller birds, but like all bats its body is small in comparison being only about 3½ inches long including the tail. Bats are not always black as they appear to be in cave lights or silhouetted against the sky. The Greater Horseshoe bat is reddish-brown above, with a greyish tinge below. Like its smaller relative, the Lesser Horseshoe, it is distinguished from other bats by a horseshoe-shaped membrane, forming a very distinctive nose flap. A Horseshoe bat has a distinctive method of flight when approaching its settling point. It turns a somersault before hanging up by its feet. The Natterer's bat also follows this method but is rarely successful at the first attempt. The Lesser Horseshoe, as its name implies, has a smaller body and less wingspan, the body being about 2½ inches and the wingspan 8½ inches.

As the Horseshoe bats are recognized by their nose flaps, so the Long-Eared bat is recognized by its long ears, so long that they take up almost half the length of the body. It often has a wingspan a little longer than that of the Lesser Horseshoe. When asleep in the normal upside-down position, it folds its long ears under its wings. It has less difficulty in rising from the ground than other species and the ears are held straight out in flight, but curled like ram's horns when crawling.

The Whiskered bat is among the smallest bats in Britain, with a head and body of only about 1½ inches and a tail rather less. It has short notched ears, but its outstanding feature is its moustache, due to a thickening of the fur which covers much of its face, almost concealing its eyes, so it is rather more than a mere moustache. Its fur is blackish, tipped with chestnut. This species is found over the greater part of England and is sometimes seen flying during the daytime.

Daubenton's bat is sometimes known as the Water bat, because of its fondness for lakes and pools where there are normally a large number of insects. Sometimes its wings actually dip the surface. It has rather large feet and the membrane is attached below the ankles. Its wingspan is about 9 inches, its body 2 inches and the tail a little less. The body is reddish-brown on the upper side and greyish-brown to whitish on the underside.

Without close examination, Daubenton's bat is easily mistaken for Natterer's bat, as it is about the same size and colouring. In fact the identification of any species depends on the sum of all its features. The Horseshoe bats are not alone in possessing nose flaps, but are the only bats in this country where the flaps are horseshoe shaped. With most bats, apart from size and colouring, two of the principal points of identification are the ears, their shape and size in comparison with the body and whether they are joined or separated at the base, and also the point of attachment near the ankle of the wing membrane. It can be seen that identification involves a thorough examination of the captive bat and the new method of identification by sounds through electronic equipment is obviously easier both to the student and the bat.

The distribution of bats through the country, like that of other animals and birds, is variable. The Greater Horseshoe, for example, is typical of the West Country. None of the bats found in caves are pure cave dwellers and are found outside caves, including the little Pipistrelle, the 'flitter-mouse' so often seen in country lanes at dusk. Evidence indicates too that many of the bats alternate between caves and dwellings of different kinds. It is possible, too, that although young are found in caves, either clinging to their mother's fur in flight or temporarily left 'at home' while the mother is on the wing, the bats find more acceptable places than caves as breeding grounds.

A great deal of research has been done on bats in caves but there is still a lot to learn about their habits. The work particularly in tracing migration involves bat ringing, which in this country was first carried out on a major scale in Devon in 1948. It is the same principle as is used for tracing bird migration. The rings differ from those used on birds, as the legs are not free like birds'. The bats' migratory journeys are not as extensive as many of the birds'. Many bats are recovered within a few miles of the cave where they were ringed and sometimes the same bat will be identified a number of times. Only occasionally do they go as far as thirty miles, but more usually range from seven to ten miles. Sometimes bats use houses as a summer retreat, or rather their roof spaces, returning to the cave in winter for their long term of hibernation. More research has to be done on the longevity of bats, but their life span has been calculated from seventeen to twenty years and even longer.

Although bat ringing has revealed much valuable information about the life and habits of bats, the practice has lately come under some criticism. If ringing disturbs bats in their hibernation sleep, they may use up too quickly their precious reserves of food stored in their bodies and so may not be able to survive until the warm weather brings a fresh supply of insects. During the 1950s and 1960s it was noticed that there had been a great decrease in the number of bats throughout Europe and, although probably not due to any single factor, it was suspected that one of the reasons was interference with their roosting places. Cavers were not entirely to blame, as bats also live in other places such as old buildings, hollow trees and belfrys. Many have lost their roosting places through demolition and clearance. However, with the considerable increase of caving activities, whether cave exploration or ringing, it is reasonable to assume that many of the animals have been disturbed. Some bats have become so used to hibernating in caves that they might not be able to survive elsewhere.

To avoid this disturbance, caving clubs refrain from visiting certain caves containing bat colonies from October to April. With large cave systems, there is less problem, as they have many passages too narrow for cavers, but the animals may be particularly vulnerable in small caves and blind passages. In 1975, there was a grave risk of a small cave at Dundry, near Bristol, known to be a habitat of bats, being lost. Two boys were trapped in it by falling rocks and had to be rescued. Members of the public asked for the cave to be filled in as they thought it a danger to children. The owners were persuaded to agree to the cave being gated instead and so the bats are safe from intruders.

Other caves have been gated for the purpose of bat conservation, but this in itself has raised a problem. In one instance, a flourishing colony of bats has reduced in numbers since their cave was fitted with a grid of 6-inch squares through which the bats could pass, but it is suspected that the presence of the gate has, in itself, frightened away some of the bats. There does not appear to be any easy solution to the problem. Not all gated caves are bat reserves, however. It cannot be too strongly emphasized that nobody should disturb a sleeping bat. Indeed the Greater Horseshoe bat and the Mouse Eared bat are now protected by law and it is illegal to handle them without a special licence issued by the Nature Conservancy Council.

There are many other forms of life in our caves, spiders, insects, worms and small water creatures, and all these are subjects for cave study. A pool of water may contain minute life which may have considerable importance to investigators. Practically every caver, whether specialist or not, has come across *Meta menardi,* the cave spider, and perhaps its near relative, *Meta merianae.* Both are frequently found as cave threshold inhabitants, that is, living just within the entrance, where there is an ample supply of flies and gnats in the foliage round the cave mouth and flying in and out of the cave. The spiders are also to be found in many man-made subterranean passages. The white bulbous cocoons containing the eggs can often be seen hanging from cave ledges immediately within the entrance. These are not the only spiders to be found in caves, as more familiar types are also present and the collector of spiders will find ample opportunity to pursue his studies of these creatures in the unusual environment of the caves.

Perhaps the most interesting forms of cave life are those which in the course of generations in caves have remained in a primitive or ancestral state, not progressing as have their surface relations, and those creatures whose bodies have adapted to cope with cave conditions. Living in the perpetual dark of the caves, some forms of life have dispensed with sight and body colour. We have no blind salamanders such as exist in foreign caves, nor the cave-adapted gecko of Ceylon, but we do have altered forms of smaller life. There are many problems to be solved concerning cave life and the reasons for adapted forms. In some instances creatures may have taken to caves to escape predators who lived on them in the open. In this way some species may have survived long after the extinction of their cousins on the surface. Experiments have shown that some cave inhabitants, although not blind, avoid light, either by retreating or burrowing, and it is this burrowing which has prevented some from being washed out of the cave in flood conditions. Many living things, although not physically adapted, change their habits within caves and become so accustomed to their new way of life that they find it difficult to exist on the surface. Sometimes the creatures themselves are difficult to find, such as worms, but leave evidence of their existence by casts.

The distribution of life in various cave zones depends in part on the type of food available. *Meta menardi* prefers the threshold zone because of the presence of flying insects which do not go far into a cave and so we do not expect to find this spider beyond the threshold zone of a cave entrance or open fissure. Worms which obtain their substance within the soil can be found well beyond the threshold zone and in fact some seem to exist entirely on non-organic matter. A number of other small cave dwellers, such as freshwater shrimps, can find sufficient feed in freshwater rock pools well inside a cave.

Since some creatures are confined to the dark zone of caves and others are not, special terms are applied to them: trogloxenes are those which do not spend the whole of their time in the Dark Zone, the bat being the obvious example. Troglophiles are those which live permanently and breed in the Dark Zone, but whose species can also be found breeding outside the zone, and lastly, troglobites are those species which live and breed only in the Dark Zone. Bats are intentional and therefore 'habitual trogloxenes', but others which get into the zone accidentally and usually do not survive there are known as 'accidental trogloxenes'.

Freshwater shrimps and springtails are among the most numerous of cave residents. As shrimps inhabit pools, they are not to be found in dry caves or the dry passages of a cave system. One genus of freshwater shrimp to be found in caves is completely blind. The differences to be found in a number of cave creatures compared with their surface companions is loss of sight, a surplus feature in the Dark Zone, and loss of colour partly or wholly, creating a transparent appearance. Sometimes there is a physical adaptation of other appendices so that feel can make up for the loss of sight. Man has come to depend almost entirely on sight to the detriment of sensitivity to feel and recognition of scent, senses which are often more acute in other animals which use them far more fully than does man. As we have seen, the bat depends entirely on sound. Changes in animal life and form are facts, but the reasons for these mutations are still a basis for study.

Animals and insects are not quite the only forms of life to be studied in caves, for certain species of fungi, parasitic on certain flies, have been found in caves and are unknown elsewhere. Among the living things in caves are bacteria, fungi and algae. Some of these can be minutely small and are known as microflora. Some of the microflora live on organic material brought in from outside or on crumbs and food dropped by cavers. These are known as heterotrophes to distinguish them from the autotrophes which live entirely on mineral substances and they themselves often provide food for heterotrophes. This microflora is capable of living in darkness, but higher plants need light for their full development and therefore are normally only visible at the threshold. That they can exist beyond this zone is shown by foliage which tends to develop round the lights of show caves. In some show caves, the growth becomes so thick that it often has to be cut back before it masks the lights.

Cavers with a leaning towards geology turn to the study of the formation of caves and the growth and forms of stalagmite. Although caves form only in certain rocks, they are of particular interest to the geologist prepared to specialize in the limestone rocks. Through caving he can penetrate into the depths of the deposits and study their formation and structure in a way not possible on the surface. Geological faults not apparent above ground may be clearly visible below and these may be a major factor in the formation of a cave. Changes in the rock may be more distinct below ground because of the lack of drift or spread which is often inevitable on the surface. Certain of the diving activities carried out by the Cave Diving Group in the Wookey Hole series were known as 'into the limestone' operations, as at a certain point upstream the rock suddenly changes from Dolomitic Conglomerate to Carboniferous Limestone with a much more distinct division than that on the surface. The lines of the jointing and the geological 'dips' and 'strikes' of the rock are very apparent in cave series.

Rocks are classified according to their origin. Limestone, which is most usually associated with inland caves, is a sedimentary rock, consisting of the hardened sediment of ancient sea beds. Igneous rocks are those formed from molten rock due to heat action within the earth's crust. Often they penetrate into the sedimentary rock, as can be seen in the limestone areas of Derbyshire. The third group of rocks in the classification are metamorphic rocks, rocks which may be igneous or sedimentary, which have undergone change due to heat or chemical action. Dolomitic Limestone is an example.

Even the familiar limestone varies from region to region. Carboniferous Limestone, the limestone associated with the same series of rocks which include the Coal Measures, is divided into various zones and sub-zones for recognition.

These zones are named after some particular fossil which they contain. Fossils, consisting of sea-shells, sea lilies (crinoids) and various types of coral, are seen to advantage in many cave walls, where water action on the rock has left them projecting and exposed. Within these fields, there is much scope for the caver interested in geology, mineralogy and palaeontology.

Linked with geology is yet another subject, that of hydrology, the study of water systems and their actions, a study which has a very practical application. In certain districts, much of our domestic water supply comes from cave streams and rivers, and knowledge of the quantity, flow and direction of this underground water can prove a valuable asset in improving the water supply and the tracing of any possible pollution. Water may travel many miles underground, often through passages inaccessible to the cave explorer. The emergence of the water from an exit cave at the foot of the hills will be known, but it is important to know where the water enters the cave system through a swallet often on the high land, for it is here that contamination may take place. Certain chemicals are placed, one at the time, in the likely streams where they go underground and a material which will react with this chemical is placed in the emergence of the water. In this way the source and sometimes several sources of the cave water can be found.

Cavers must know about the course of underground streams for their own sakes, as in flood conditions their escape may be cut off and then they must know which passages are likely to flood and which will remain dry. The rescuers, too, must know the principal sources, so that these may be diverted, if necessary, as was the case in 1951 when I was involved in a rescue as the result of flooding at Ogof Ffynnon Ddu. I and others spent twenty-eight hours on a mountain, assisted by coal-miners and soldiers, diverting the principal source. As a result, we decreased the amount of water flowing into the cave. This, and an improvement in the weather, enabled the two trapped cavers to walk out safe and well.

The mapping of underground streams involves a good bit of surveying as does the discovery and exploration of new caves and passages, which have to be recorded and surveyed in some detail. Some cavers specialize in underground surveying and map-making. Many of the instruments used for surface surveying, such as theodolites and dumpy levels, are too cumbersome for underground work and the space too restricted. The traditional instrument is the miner's dial, but various improvised instruments such as converted aeronautical compasses and home-made equipment have been used.

One of the problems is that the surveys usually have to be done in exceptionally poor light and the lenses of telescopic instruments are less effective in such conditions. Much of the surveying is therefore done by unaided vision sighting and, because of long narrow passages, the open traverse method of survey has to be used rather than the more accurate and easily checked method of triangulation, normally used on the surface. The open traverse method can be checked by resurveying back to the starting point or by checking every sighting in reverse, but this is long and tedious.

Although it is not vital in many caves for a cave survey to be as accurate as a surface survey, it can be important where the object is to ascertain the thickness of rock and the exact place to break through to join up with a neighbouring system and it might be necessary to break into a system for rescue purposes. Many cave clubs grade their plans according to their degree of accuracy.

Most caves have corresponding surface features such as swallets or sink holes. An underground survey should link these features and show the relation of the cave to the surface. It would not be worth spending years digging a swallet in search of a new cave to find oneself in a known cave. Sometimes it is thought expedient to have more than one entrance to a cave system, particularly a system that runs into miles, such as the alternative entrances to Ogof Ffynnon Ddu, not only to reduce the waste of time in covering old ground to explore new, but for speed of rescue if necessary.

Without accurate surveys easy ways for the public to enter the caves by tunnels, as at Dan yr Ogof caves, would not have been possible. Questions of ownership of caves or possible danger to surface development all require underground surveys linked with surface features.

Some advance has been made in mapping caves from the surface by means of sound and electronic equipment. Such equipment has been used on known caves for comparison with actual underground surveys with good results. It is known as geophysical surveying and various types of geophysical equipment have been used. The surface of the ground above the cave system is divided into a grid of squares and soundings, taken at each intersection. One favourite type of geophysical survey is by electrical resistivity. As its name implies, it is the passing of an electric current by metal stakes at each intersection of the squares and measuring the resistance of the underlying strata by the voltage drop. Underground air spaces have no resistivity.

Much of the pioneer work was done with this method by the late Professor L.S. Palmer on Mendip with the assistance of members of the Bristol University Spelaeological Society and others, using the known cave system of Lamb Leer and other sites. Except for some variation, where a system was small and close to the surface, the results showed close relationship to the underground plans. The system had been used by geologists over a long period for locating changes in rock resistivity, denoting different strata and subterranean geological changes, including mineral identification. We owe much to Professor Palmer's activities for application of this method to cave detection.

The method was used to trace an ancient mine in the Southmead area of Bristol, known as Pen Park Hole. The mine had been capped for years, but had gradually become covered with soil and lost. Housing development had been taking place in the area for some years and it was advisable to know the exact location of the underground cavity before extending the development. Professor Palmer's use of the geophysical apparatus not only located the old shaft within a foot, but traced the extent of the underground system.

The relationship of a cave system to surface features has also been established by radio control between the explorers and a surface team. The stronger the reception, the nearer is the transmission and so the thinner is the rock cover between underground and surface teams. In this way can be established not only the relationship of a cave to the surface, but also the most economical position for driving an additional entrance, if required, a method which was used at Wookey Hole Cave, Somerset.

Perhaps the most popular cave study or hobby is that of cave photography. It soon becomes apparent to the newcomer to caving that vistas of cave passages and the spectacular columns and curtains of stalagmite are unique and fascinating subjects for photography. Since caves are in perpetual darkness, lighting is all important and many unusual effects can be obtained by strategic placing of lights. Nowadays there is quite a sophisticated range of electronic flash, flashbulbs and other light sources. By using several points of light, silhouetting some formations and illuminating others, a great variety of interesting and artistic photographs can be taken. Many are published in the various cave club journals and from time to time exhibitions of cave photographs are arranged by caving organizations and these are always well worth seeing. I rather think some sort of scale in a picture is essential, whether it be a matchbox or a person, because cave scenery does not relate to familiar surface surroundings and the viewer has no means of telling whether the formations photographed are gigantic or tiny. There are several articles on the technique of cave photography in some of the general caving publications listed in the bibliography.

Some of the earliest cave photographs were taken by the late Harry Savory, with a large wooden plate camera —cumbersome and awkward equipment to get through small cave passages; time consuming too, as many were taken with long exposures. The late H.E. Balch, one of the great pioneer cave explorers, told me how he had to spend what seemed to be an infinitely long period sitting on a boat in the middle of the River Axe in Wookey Hole, Somerset, while Harry Savory took a photograph of what was to be one of the principal advertising illustrations of the show cave for many years. His photograph of the Witch of Wookey was another well-known illustration for many years. Harry Savory still used the old type plate camera until his death in 1961, although most of his later pictures were of natural history rather than cave scenes.

My own feeble efforts at cave photography were made

many years ago with a box camera, aided by liberal piles of magnesium powder placed at strategic points, such as behind stalagmites, to give relief to the picture. Candles and torches were extinguished, and the camera shutter opened in the darkness. Members of the team placed close to the magnesium would then light the blue touch paper and either duck out of the way or tear across the rock-strewn floor in the dark, hoping to get out of the picture by the time the flash powder ignited. Often this dash was of no avail, as the touch paper, dampened by the cave atmosphere, went out, or a heavy splash of water from the ceiling landed fairly and squarely in the middle of one of the heaps of magnesium powder. Magnesium ribbon would occasionally be used but it was not always easy to exclude the operator from the picture. Usually with the use of magnesium, a heavy pall of thick smoke would hang about the cave, making too much haze for further photography for some time.

Although we always did our best not to litter the cave with debris, it was not always easy to avoid losing the odd candle stump, dripping candle grease on rock, leaving a spatter of magnesium and occasionally, but not intentionally, spilling spent carbide. Modern equipment is much improved in this respect, leaving little or no debris behind, and our caves should be cleaner in consequence.

Photography is a useful adjunct to all cave studies, whether they be biology, botany, or geology, but perhaps it is most important for cave records. Sometimes caves are destroyed by quarrying, sometimes they are sealed for safety reasons, sometimes they are changed by roof falls, and sometimes opening a cave to the public means destroying some of the natural features. A cave at Penderyn, South Wales, and many others now only survive in photographic form, as they were discovered during quarrying and disappeared for the same reason. There are many photographs of the 35-foot —commonly known as the 40-foot—waterfall in Swildon's Hole, Mendip, but one day, when fortunately no cavers were present, the rocks gave way and converted the fine fall into a slope. For over half a century, cavers had belayed their ladders at the top of this fall and it seemed inconceivable that what appeared to be a solid rock platform would one day collapse into the lower chamber. In other cases rock movements have only become apparent when photographs taken at different times are compared. Climatic conditions vary and cave streams may even change their courses for one reason or another, and here again, photographic records are most valuable.

Over the years, research into caves and their formation has led on to the study of subjects associated with caves and with cavers themselves. From time to time, we hear of cavers undergoing duration tests in caves, spending long hours, days and even weeks in the dark alone. Although it has an element of record breaking, such occasions are usually carefully monitored by the medical profession, often by doctors who are themselves cavers, and a great deal of information has been accumulated about the reaction of the human body under such conditions, including the effect of exposure, diet and isolation. Such information is valuable in safeguarding the health of the caver and also provides useful data for other spheres of activity.

Cave studies are not new. The study of biospeleology or cave life, for instance, was being carried out as early as 1701. The increase of caving activities and the advance of scientific aids in recent years, together with the results of surface study, have resulted in a considerable growth of knowledge in all fields. However, there is still a great deal to learn and the beginner need not be daunted by the wealth of existing material. In fact, it would seem that new knowledge brings new problems to solve. Each field of cave study has become so extensive that no one person can hope to learn all there is to know about every aspect of caving. On the other hand, if you want to become a serious caver, it is as well to know at least something about each study, if only to know how to avoid unnecessary destruction to the caves themselves and to the things they contain.

FIVE

Bone Caves

FROM EARLIEST TIMES, men have lived in caves and have found them highly desirable residences, providing ready-made protection against the weather, easily defended against wild animals and human enemies. A dry cave or rock shelter must have been a great find even if the man had to oust an animal occupant. Settled in their cave home, the family would soon scatter their rubbish around them, food bones, broken crockery, lost ornaments and discarded tools. Sometimes even their dead would be buried there. If the succeeding layers of cave deposits are removed one by one, their contents can be read like the pages of a book, beginning at the end of the story with the surface layers and working down to the beginning with the lowest level of the earliest deposits.

Unfortunately, unlike the pages of a book, the various deposits do not lie evenly. They conform with the irregular configuration of the cave floor, vary in thickness, and are not all evenly distributed. Some of the pages have been torn and mutilated, or are missing in places. Sometimes the deposits have been disturbed by burials, sometimes objects have fallen a considerable depth down gaps between great boulders, and sometimes burrowing rabbits have moved material from one level to another. Often it is not easy to distinguish one layer from another and one particular band may have been laid down over a long period and contain artifacts of different ages.

The chronological sequence of the deposits and their relationships is known as stratigraphy and this is the basis of all archaeological excavation in caves and on the surface. In both instances, stratigraphy shows the sequence in which the deposits are laid down, but the origins differ. In a surface excavation, much of the stratigraphy below the top soil may be due to man's own interference, the levelling of former buildings, plant destruction and dust deposit and old rubbish tips. In a cave, the subsoils owe their origin to various elements responsible for the cave's formation, debris from the roof and walls, silt and material brought in by cave streams and the formation of stalagmite. The cave excavator needs to know about cave formation and has to be something of a specialist in those branches of geology and hydrology relevant to caves. He must know how to distinguish the deposits by the fossils they contain and not only be able to identify the remains of modern animals, but also those which have lived in this country in the distant past, such as early rhinoceros and elephant.

Most of our knowledge of the latter part of the Old Stone Age has come from cave deposits and the British caves have been reasonably productive in finds of this period. We do not have the finely decorated bone material of France or cave paintings, but our caves have produced quite an appreciable number of flint implements and other artifacts as well as the bones of animals who lived here in the Old Stone Age.

Probably the first archaeological excavation to be carried out in a British cave was at Oreston near Plymouth in 1861, when rhinoceros remains were found. Many cave systems have been broken into during quarrying operations and this cave was discovered while quarrying for stone for the breakwater at Plymouth. Long before this discovery, bones had appeared in various parts of Britain, including bones which had been described as elephant. Many arguments raged around the question of their origin. People believed that the world had been created in the same image as they knew it, with the same distribution of animals and the same climates. How did elephants get from the tropics to Britain? It was not realized that the mammoth (which had been mistaken for elephant) was an extinct species which lived on the fringe of ice-bound territories and was not a tropical animal. It is now known that Britain had a range of climates from tropical to intense cold.

There were all kinds of theories to explain the presence of these strange animals so far from their usual habits. The most popular was the 'Deluge' theory which stated that the Great

Flood had washed these animals from their native lands to deposit them in the caves and crevices of Britain. This explained the dismemberment of the carcases, so that only part of the animal is found. One fantastic theory was that the caves themselves owed their origin to the decaying animals. The rocks, converted into mud by the Great Deluge, contained the animals' remains and when these decayed, the gases of putrefaction caused bubbles to form within the drying mud, so that caves were formed. Some people even went so far as to suggest that the size of a cave was in proportion to the carcases it contained, in spite of the fact that there are many caves without bones. According to these early theories, the gas sometimes escaped through weak points of the earth's crust before the bubble reached its full size, thus making passages and fissures leading from the bone chamber.

These theories seem ridiculous to us today, but it must be remembered that practically everybody believed that the Deluge covered the whole of the earth and that certain of the rocks owed their origin to it. We now know that the Bible Flood story is derived from much earlier writings of Sumerian and Babylonian origin. Evidence of extensive flooding in the Tigris–Euphrates area was found in the form of flood deposits by Sir Leonard Woolley during his excavation at Ur and other sites and the extent of this flood could actually be ascertained by trial holes in the area. This extent was much of the Sumerian and Babylonian world. However a literal interpretation of the Book of Genesis was almost absolute, not only among ordinary people, but among the most eminent scientists and thinkers of the age. There was a tendency to reject any explanation of natural phenomena which did not fit in with this interpretation and since the Deluge was believed to have covered the whole world, including Britain, it was inconceivable that there was not some evidence of it in the rocks of this country. There seems little doubt that this blind adherence to 'Mosaical Geology', as it has been called, must have had a retarding influence on the advancement of geological knowledge of that time.

It was the Reverend William Buckland, Professor of

Geology at Oxford, later to become Dean of Westminster, who began to open the way to a better understanding of the significance of these strange animal remains in our caves. It was in 1821 in Kirkdale Cave, near Kirby Moorside in Yorkshire, that Buckland established the fact that animals such as the hyaena were once living in this country as well as other now extinct animals. His opinion was either the mass of hyaena bones there was brought in by the 'Diluvian waters' or the cave had been a 'hyaena's larder'. He quickly concluded that it was a larder. The bones had been polished by later generations stepping on them, the corners of the passage had been worn smooth by constant rubbing of the animals' hair as they passed in and out of the cave. There were also bones of rhinoceros, hippopotamus, deer, horse, ox, fox and water rat, some of which would have been too big to have entered the cave as living creatures and must have been dragged in as parts of carcases. He described the animals as 'antediluvian'. A number of bones had been gnawed by, as he later proved, hyaena teeth. To prove his theory, he kept a live hyaena in captivity and studied the marks it made on its food bones and the rubbing of its hair on the sides of its den. One of the most important finds in the cave in support of Buckland's conclusion was the vast amount of what was described as 'album graecum', the fossilized excreta of hyaenas. This was compared with that produced by the live hyaena and found to be similar.

W.D. Conybeare depicted the Professor in Kirkdale Cave, surrounded by live hyaenas, and accompanied his cartoon with the following verse:

> But of all the miraculous caves,
> And of all their miraculous stories,
> Kirby Hole all its brethren outbraves,
> With Buckland to tell of its glories.
>
> Ages long ere this planet was formed,
> (I beg pardon—before it was drowned),
> Fierce and fell were the monsters that swarmed,
> Roared, and rolled in these hollows profound.

I can show you the fragments half-gnawed,
Their own album graecum I've spied,
And here are the bones that they pawed,
And polished in scratching their hide.

Mystic cavern! thy clearness sublime
All the chasms of history supply.
What was done ere the birthday of Time,
Through one other such hole I could spy.

Although Buckland had proved that such animals had lived in Britain and could not have been carried here by flood from far-off lands, he clung tenaciously to the belief in a world-wide deluge and gave the word 'diluvian' to certain rocks which he believed had been laid down by the Noah Flood. This word has survived in the geological sense, meaning those sands and muds laid down by our rivers. He called the rocks which he believed had been laid down before the Flood 'antediluvian' and those laid down after the Flood 'post-diluvian'. In fact he was a firm 'diluvianist' as those who held belief in Noah's Flood were called.

Because of these rigid beliefs, the idea that man could have existed in this country in pre-Flood times was untenable to Buckland and so any human remains brought to his notice were classified as 'post-diluvian', a view held by other scientists as late as 1850, even though in many cases the association of the human bones with 'antediluvian' fauna was overwhelmingly apparent. It was a case of blinding oneself to the evidence if it conflicted with established views. All kinds of devious explanations were produced to account for the inconsistencies. A good example was Buckland's investigation of Goat Hole, Gower, better known as Paviland Cave because of its proximity to Paviland Farm, following human remains found there in 1822. The bones had been found in close proximity to those of mammoth and other contemporary animals. The identification of ancient animal bones had been well established by that time. The fact that the human bones were associated with 'antediluvian' fauna was not acceptable to Buckland and he decided that they were of Romano–British origin, because there was a small pre-Roman Iron Age hillfort on the cliffs above. Although numerous articles fashioned from bone and ivory from extinct animals were found with the Paviland skeleton, apparently this did not influence Buckland's opinion in the least. These, he said, must have been made from fossil ivory and bone in Romano–British times, although this would have been a difficult feat, since ivory and bone which have been buried for a long period lose their character and tend to crack.

The skeleton found in Paviland Cave was not complete and the skull was missing, as is often the case since the thin bones of the hollow skull are particularly vulnerable to crushing and disintegration. The presence of a skull would have been of great assistance in sex and age identification, but nevertheless Buckland rightfully identified the remains as a man in the first instance, but later changed the identification to that of a woman, because of a necklace of pierced wolf and reindeer teeth. It seems strange that Buckland should have allowed himself to deviate from his own sound anatomical knowledge because of a necklace. This might have been exclusively feminine wear in Buckland's circle, although not wolf and reindeer teeth, but even in his day many primitive tribes wore shell and bone necklaces, and even nearer at home he must have seen gypsies and sailors with earrings. As the bones were stained with red ochre, the skeleton became known as the Red Lady of Paviland. It was not until 1911 that the French prehistorian Cartailhac confirmed that the remains were that of a young man of Old Stone Age date, and this is still the earliest known cave burial in Britain. Painting the dead body with ochre was a general feature of the late Palaeolithic (Old Stone Age) burials. Unfortunately the title of the skeleton as the 'Red Lady of Paviland' had been too firmly established over the years to change it to the 'Red Man of Paviland' and would, at that stage, have been confusing.

It was in 1822 that Buckland incorporated his conclusions in a work he called *Reliquiae Diluvianae*, 'Flood Relics', but not many years were to elapse before the belief that certain deposits had been laid by the Deluge was discarded and their origin attributed to the results of the Ice Ages. Animals not only lived during the Ice Ages, for unlike the traditional Flood the ice did not cover the whole earth, but in the periods between the glaciations when the climate could be tropically hot. These periods we call 'interglacials', and the periods of extreme cold we call 'glacials'. Most of the Old Stone Age material found in caves belongs to the period of the last Ice Age or the Late Palaeolithic, if we use the archaeological term, or Late Pleistocene if we prefer to be geologists. The mammals associated with such deposits are called Pleistocene mammalia, but the man-made material is always known as Palaeolithic artifacts. Some caves produced warm fauna of the last interglacial period such as Victoria Cave, Settle, excavated by Boyd Dawkins and others, and Minchin Hole, Gower, where Jim Rutter and I excavated for many years in deposits which had produced straight-tusked elephant, forerunner of the mammoth, and soft nose (i.e. hornless) rhinoceros, and some of the material we found is to be seen in the museum of the Royal Institution, Swansea.

The question of extinct animals being contemporary with man remained an unsolved problem until it was decided to put it to the test by controlled cave excavation. There had been a number of discoveries of Pleistocene fauna in proximity with Palaeolithic implements, but many prehistorians were not convinced. There was always the possibility that bones or artifacts of later date could have been buried and so found at a lower level. If a cave is excavated properly, there should be evidence of any such burial. The cut through the deposits should show the distinct sides of a pit and the contents of the pit will have no stratigraphy, the fill being a mixture of all deposits disturbed. In fact, if it is an ancient burial, the level at the top of the pit disturbance will indicate when the burial was made.

It was decided that one way of proving if bones and artifacts were contemporary was to find them in a situation which it would have been impossible to reach without leaving very clear signs of disturbance. What better deposit could there be but one covered with a thick layer of stalagmite. Many caves contain such a layer in their deposits. At Minchin Hole, Gower, the remains of a deposit 8 inches thick can still be seen adhering to the walls, once part of a complete layer which extended from wall to wall—until a local squire came along in the last century in search of curiosities. He had it broken by explosives to retrieve the

Carreg Cennen Castle, under which Owain Llawgoch
is said to sleep

The mystery of
Ogof Ffynnon Ddu
—the human bones
as they were found

The Witch of Wookey

The Main Chamber, Gaping Gill

Abseiling in Alum Pot

In Long Churn Cave

Alum Pot

Stalagmitic columns
and flowstone in
Stump Cross Caverns

Climbing in Nettle
Pot

Inside Eldon Hole,
at the top of the
80-foot pitch

West Chamber,
Oxlow Cavern

tusks of elephants and the bones of rhinoceroses from underneath the stalagmite layer.

The fact that explosives had to be used at Minchin Hole shows what tough material stalagmite is. It is very easily broken and brittle in stalactite form, but when it forms a stalagmite floor over cave deposits it is difficult to penetrate, certainly a continuous seal of stalagmite is a guarantee that the material below it has not been disturbed.

There is no doubt that the publication of *Reliquiae Diluvianae* caused increased interest in cave excavation. Within a few years of the publication, the Reverend J. MacEnery started excavations in Kent's Cavern, then known as Kent's Hole, Torquay, on which he worked until he died in 1841. It may seem strange that so many reverend gentlemen became involved in a subject which seemed to threaten some of the religious beliefs of the day, but education and leisure at that time were the prerogatives of the wealthy and professional classes. MacEnery found flint implements and extinct animal bones in the same context, but other experts were still not convinced.

In 1858 Brixham Cave was excavated by the Royal Society and the Geological Society and it was there that ideal conditions were at last found, an undisturbed stalagmite layer under which were flint implements, together with bones of mammoth, woolly rhinoceros and hyaena. Other caves had been producing similar information, but the carefully controlled excavation at Brixham established once and for all man's association with Pleistocene fauna. Interest in solving the problem was not confined to caves. Flint implements with extinct animals had been found on old river terraces at Abbeville and Amiens in France, some years before the Brixham discovery, by Boucher de Perthes. His long struggle to convince his colleagues that man-made flint implements were contemporary with extinct animal bones was brought to a triumphant conclusion by the Brixham discovery.

Boucher de Perthes was as eccentric a character as many of the early archaeologists. His name Jacques Boucher de Creve Coeur de Perthes, 'Butcher of the broken heart', has no connection with the fact that he published the love letters reputed to have passed between Pauline, sister of Napoleon, and himself, many years previously. He was said to have descended from a well-established line and his mother was believed to be a descendant of one of Joan of Arc's family. He had been described as a pamphleteer, poet, playwright, novelist and politician. He certainly dabbled in a number of things, not always successfully. A lot of his writing was never published and his political career was doomed from the start. How could such a radical expect to be elected with his mad ideas such as world peace, old age pensions and the abolition of capital punishment? Unfortunately many of his writings were destroyed by his family after his death and such publications as existed were withdrawn. Perhaps they were not satisfied with his will, for instead of leaving his fortune to his family, he left it to the poor working women of twenty French towns.

Work was not continued in Kent's Cavern until 1865, when excavations were carried out by Mr W. Pengelly, for the British Association. For some years the work produced large collections of flint implements in similar circumstances to those previously found by the Reverend MacEnery. The name Pengelly is today promulgated in the William Pengelly Cave Research Centre, founded in 1962. The Centre at Buckfastleigh in Devon is used for the furtherance of cave study and conservation.

In 1859 William Boyd Dawkins and the Reverend J. Williamson commenced the excavation of the Hyaena Den at Wookey Hole, producing flint implements, and bones of hyaena, mammoth, woolly rhinoceros and other animals. It was at this time, too, that many of the famous caves containing Pleistocene remains were excavated in Gower. The work at the Hyaena Den was taken over by H.E. Balch, the pupil of Boyd Dawkins. He later transferred his activities to Badger Hole, where he worked for many years. I remember working at the Badger Hole under the guidance of H.E. Balch, with cavers who are still friends of mine today. We would laboriously fill buckets, which were emptied into a rocker sieve if the soil was friable. If it was clay, as it usually was, it was emptied on to a table, at which the 'grand old man of Mendip', as he was called, sat and he would proceed to cut the clay into slices with a wooden slicer, skilfully extricating any object he came across. Bone fragments would be placed in numbered boxes according to the area being excavated, each specimen to be marked individually. The debris left over was emptied into an iron box on an endless cable which was wound out of the cave, where it ascended to the debris heap. When it reached the heap, a pole released a lever on the box, opening the bottom. The incline was such that the box, when empty, travelled back to base of its own accord.

While Professor W.J. Sollas was carrying out further excavations at Goat Hole, Paviland, in 1912, his attention was drawn to some ochre streaks in the Gower cave known as Bacon Hole. No cave paintings such as those found in France and Spain have ever been recognized in Britain, although similar flint implements have been found. Shaft straighteners or *batons de commandement*, similar to examples associated with such paintings, were discovered at Gough's Caves, Cheddar. Sollas thought that the streaks in Bacon Hole were not unlike streaks of primitive art found in the French and Spanish caves. His opinion was published in the press and a scathing letter was printed in a national newspaper from one of its readers, who said he knew the origin of the marks—a boatman, painting his boat near the cave, had wiped his paint brush on the cave wall. Sollas must have been well aware that this was an unlikely explanation, as there is no beach of any consequence at Bacon Hole and certainly not one where a man was likely to keep a boat. Seaweed-covered rocks are everywhere, entirely covered at high tide. By the time a boatman had climbed into the recess of the cave where the marks are found, it is more than likely that his brush would have dried out, even if he had thought the effort was worth the trouble.

During the Swansea Conference of the British Speleological Association in 1939, Marett, former assistant to Sollas, then a well-known excavator noted for his discoveries at a cave at St Brelade in the Channel Islands, mentioned these streaks to the famous Spanish prehistorian, Bosch-Gimpera, who examined the marks and gave as his opinion that had he seen them in a Spanish cave, he would have been convinced that they could have been primitive painting. It was hoped at

that time that a more thorough examination could be made
and if, as expected, the streaks proved natural or modern, at
least the story would be given a scientific burial. From
photographs taken at intervals, it was established that the
streaks were not always in the same position, but this fact
alone does not prove that the marks are natural. I know of
one historical painting on a castle wall which only appeared
piecemeal according to which part of the wall happened to be
damp, until a complete restoration was carried out. It is
believed that ochre, without a binder, will not remain on a
cave wall indefinitely. In fact, an organic binder must have
been used in prehistoric cave paintings. Traces of the binder
have now disappeared and the colour is often now secured
only by a calcite covering. The paint daubing story appears in
more than one version and required more authentic evidence.
On the assumption that modern red paint contains other
ingredients than ochre, it would seem that simple analysis
would lay the ghost of Bacon Hole for ever. During conver-
sation at the Speleological Association in 1939, Marett was
asked why Sollas had never challenged his critic. He replied
that Sollas was an old man at that time and did not wish to
become embroiled in a lengthy argument. Many people have
searched for traces of prehistoric cave paintings in this
country, but all known caves are constantly visited and likely
marks, however faint, would hardly be missed, so it is
unlikely that prehistoric art in Britain ever existed on cave
walls.

   Throughout the ages, there has been no period when our
caves were not inhabited by man from time to time and they
have always been the dens of animals. Such an animal den
was found by Beard at Banwell, Avon, where he discovered
bones of cave bear, cave lion, reindeer, wolf, hyaena, bison,
mammoth and woolly rhinoceros. There were so many of
them that after removing the best specimens, he packed the
remainder in patterns against the walls of the cave, in the
way that human bones are often arranged in an ossuary.
Although he carried out his excavations in the 1820s, his neat
arrangements of bones have survived there to this day.
Fortunately, as the cave is in private grounds, it has escaped
the present-day vandalism and a small dedicated group of

cavers have made it their concern to preserve this unique collection on the site.

The period which followed the Palaeolithic was the Mesolithic or Middle Stone Age. The mammoth and rhinoceros had long since gone and it was during this period, between 6000 and 9000 B.C., that the melting ice caps caused a rising in sea level and separated the British Isles from the European continent. During the Mesolithic period, Britain and the animals which had come northwards in the wake of the melting ice began to assume something of the pattern we know today. Some of the best evidence of these changing times came from Creswell Crags, Derbyshire, which produced not only Mesolithic flint artifacts but also material of the late Palaeolithic, named Creswellian. Neither our Mesolithic nor our Palaeolithic material compares with the perfection of that found in Europe. We have nothing to compare with the wealth or standard of the artifacts found in the great Mesolithic cave of Mas d'Azil in France, but neither do we have any cave that compares with the grandeur and size of that cave, which is so large that it is used as a highway tunnel.

One of the major interests in our cave archaeology is the development of native cultures, such as the inferior Creswellian compared with the advanced cousin, the continental Magdalenean culture, and our own Mesolithic cultures compared with the superior continental cultures of the same period. This idea of a poorer but interesting native culture also applies to some extent to later periods represented in caves.

The Mesolithic people were succeeded by the first agriculturists in this country, the small long-headed people of Iberian stock, the New Stone Age or Neolithic people. Their round-bottomed pottery bowls have been found in many caves. Soon after 2000 B.C., these peaceful people were overrun by the belligerent stocky bullet-headed people who represent the beginning of the Bronze Age, as they were the first to introduce the use of metal to Britain in the form of copper–tin alloy—bronze. They also had a liking for gold, such as the hollow gold bracelet and ring found together with a number of bronze weapons and implements in Heathery

Burn Cave, near Stanhope, County Durham, by Canon Greenwell (known to anglers for the Greenwell fly) in the later part of the last century. A solid gold bracelet of this age was found by Gwilam Harris at Bracelet Cave, Ebbor, Wookey Hole, in 1955.

The end of the Bronze Age was marked by the arrival of the pre-Roman Iron Age people about 450 B.C., who brought with them iron implements and weapons. A number of caves, such as the cave at Bishop Middleham, near Durham, and Harborough Cave near Brassington, Derbyshire, have produced evidence of their occupation. There is quite a lot of evidence of people living in caves during the Roman occupation. Kinsey Cave, Giggleswick, Victoria Cave, Settle, Sewell's Cave, near Buckhaw Brow, Yorkshire, Thirst House, near Buxton, Derbyshire, and Wookey Hole, Somerset, proved particularly rich in remains of the period.

I excavated such a cave myself from 1938 to 1950. The cave was above Dan yr Ogof caves in the Swansea Valley and was known as Ogof yr Esgyrn (Cave of the Bones) because of the large number of human and animal food bones found there. It contained not only bones of the Roman period but also some Bronze Age material, including a gold bead, bronze awls, bronze rapier and a bronze razor. By far the larger quantity of material belonged to the Roman occupation.

Our first task was to clear the floor of boulders and a couple of recently dead sheep. At the same time, the cave was surveyed and the floor area divided into yard squares. The method was to excavate one square yard at a time, noting the stratigraphy and position of artifacts in each square, and marking and recording each article and bone with the square number.

During the clearance and survey, a deep pocket of sand was found against the cave wall and this pocket proved to be the principal burial area in the cave. Most of the other bones were found within the top foot or so of the deposits and often immediately below the surface. The Bronze Age rapier was found sticking upwards in a rabbit burrow, with the bones of a rabbit round its point. The burrowing had stopped when the animal apparently became impaled on the point of a weapon, probably well over 2,000 years old at the time. All the Bronze Age material was dated to between 850 B.C. and 1050 B.C.

All the human bones at Ogof yr Esgyrn appear to have belonged to the Romano–British period, there having been no less than forty bodies, and these were of fourteen adults, twenty younger people and the remainder too fragmentary for age identification. It was not possible to be precise about the actual number because of disturbance over 2,000 years. Eight of the children had a narrow age range of six to seven years at death, based on radiology examination of the teeth. The metal objects and pottery indicated two separate periods of occupation during Roman times and the burials were of the earlier period prior to A.D. 180.

An intriguing point about the Romano–British artifacts was that, in addition to coins, rings and brooches, two articles were not what one would have expected to find in a cave community living in fairly primitive conditions. One was a seal box and the other a small steel yardarm, graduated to weigh up to 20 librae (about 6.5 kgs). The arm had been broken at some time and repaired by enclosing the break in bronze splints. A seal box is a contrivance for sealing dispatches. What was a family living in cave conditions doing with a seal box and a steel yardarm? Had they fallen from better days? were they fugitives, or were the articles loot? Whatever the reason, the occupations of the cave were certainly not temporary.

Most of the work of actual excavation, as is usual with all archaeological sites, was carried out with trowels and small tools. After the excavated material had been carefully searched on the site, it was taken to a large rocker sieve in the daylight outside the cave, in case anything should have been missed. Excavation at all times is a very slow process and much of the time is taken in recording and measuring. Not one item must be overlooked and many hours might be spent joining pottery and bones together. On one occasion, we found a very small fragment of what might be mud-covered cloth, which seemed to have coloured and white threads. Could this be a fragment of clothing once worn by the cave dwellers? It could have been too valuable to try to disentangle at the cave, so it was immediately packed in a moisturized container before it could dry out, and dispatched to the National Museum of Wales. There the decision had to be made whether to treat it or forward it to the British Museum laboratory. It was decided to make an attempt at the National Museum and the suspected fabric was placed between glass sheets which were gradually tightened together to squeeze out the moisture. Imagine our surprise when some days later we were presented with a small knot of string used in our survey and a small length of coloured thread from the belt of one of our lady helpers!

Many caves contain more than one occupation, but it is very rare when all periods are represented in one cave. Ogof yr Esgyrn had Bronze Age as well as Romano–British deposits. Victoria Cave, Settle, Yorkshire, contained Pleistocene remains, as well as Mesolithic and Romano–British relics, while Kent's Cavern, Torquay, produced Iron Age material as well as Pleistocene remains.

Cave excavation is meticulous and nobody should undertake it except under the supervision of a competent Director of Excavations, for it must be remembered that like many areas of cave research, excavation is destruction and a bad excavator might be asked to account for his actions. Decisions may not always be easy to make. Sometimes it may be necessary to deviate from the recognized procedure. At one place in Ogof yr Esgyrn, a slab of rock had spalled from the ceiling and stood in monolithic fashion between ceiling and floor. After some discussion with cavers who specialized in the use of explosives, it was decided to blast it. Starting from the top of the rock, it was broken by minor blasts until only the stump was left, which was then cracked into several pieces by a sledge hammer. When the pieces were lifted out, two Roman coins were found pressed down by the rock and under the bottom one was a round disc of linen, the remains of the bag which had once contained the coins.

Conservation of material is important on any archaeological site and, although the final conservation may have to be done in a museum laboratory, the excavator must be knowledgeable in emergency preservation. The wrong treatment may cause irreparable damage or the methods used may even make permanent treatment difficult. Any coin specialist or numismatist will tell you of coins where valuable mint marks have been obliterated by drastic methods of cleaning.

It would be impossible in a book of this size to enumerate all the bone caves any more than it would be possible to list every known exploratory system, and it is possible to refer only to comparatively few. Primitive men looked for a dry cave with an easy open entrance where they could live within sight of daylight, preferably with a rock platform outside where they might live and work in good weather. A river within reach for water and a steep cliff below for defence were other desirable features. As a general rule, ancient men and animals did not like the dark frightening depths and tortuous passages of exploratory caves. Bone caves are often small, no more than a single chamber and sometimes a rock shelter or the entrance chambers of an emergence cave system.

Sometimes such caves are found in groups, particularly where they are in gorges, as in Cheddar Gorge, Somerset, as in Ebbor Gorge of the same county, and in the Manifold Valley on the Derbyshire–Staffordshire border, but many are isolated.

The collections from a single cave are not always to be found in one museum and this can be annoying when one may have to travel distances as far apart as Scotland and Cambridge to see material from one particular cave. Finds from the same cave may be scattered over fifteen different places or, what is worse, in private hands. It is a pity that the products of any cave, with perhaps the exception of articles of unique national interest, cannot be housed near their place of origin.

SIX

Caves in Legend and History

To lay the lorn spirit, you o'er it must pray
And command it, at length, to be gone far away,
And, in Wookey's deep hole,
To be under control,
For the space of seven years and a day.
If then it return, you must pray and command
By midnight,
By moonlight,
By death's ebon wand
That to Cheddar Cliffs now it departeth in peace,
And another seven years its sore troubling will cease.

If it return still
As, I warn you, it will
To the Red Sea for ever
Command it: and never
Or noise more or sound
In the house shall be found.

Jennings, *Mysteries of Mendip,* 1807

CAVES HAVE ALWAYS been places of mystery to the uninitiated and are old in legend. Men never knew what they might find if they penetrated the darkness which went on and on into the depth of the hills. Cave legends are world-wide, but here we are only concerned with those of Britain. Some caves have historical associations but often in a broad sense, as we cannot be sure that the kings and outlaws really did hide in the caves bearing their names. In many cases legend and history cannot be separated.

Variations of certain cave legends appear in several places in Britain, especially the story of the dog who eventually emerges miles away, the wandering musician whose music is heard far into the cave, and the sleeping hero with his name changed to that of the local traditional hero.

The story of the sleeping hero differs in all respects from other cave legends, except perhaps for the presence of treasure, which is always subsidiary to the theme. Stories from different localities change only in detail, which indicates that the sleeping hero legends originate from the same source. It is a story of hope that when all human effort fails against overwhelming odds, the superhuman power of the hero will come to the rescue. This hero conveniently lies in storage preserving his youth and strength until he is needed and in what better place than a cave, a dark, secret place, hard to find? He is living and near at hand, but cannot be seen because he is sleeping in some hidden cave, away from prying eyes. Those who do find him, live just long enough to testify that he does exist, or if they do survive, they become crippled or mad and cannot find the cave again.

Sometimes there is treasure in the cave, guarded by the sleeping hero and his men, but this, I feel, is an addition to the story. More important is the horn or bell and the sword. The intruder makes the mistake of sounding the horn or ringing the bell before drawing the sword. The horn or bell is reserved for a call for help, but only after the sword has been drawn. This is a test of the hero or the coward. In fact, the attack on the visitor who makes this mistake would appear to be one of contempt for his cowardice, rather than anger at an attempt to steal the treasure, when treasure appears in the story.

Some mythologists suggest another slant to the tale. The hero and his men had been granted immortality on their deaths and so sleep once again in the womb of the earth, the cave, to be reborn when their people need them.

One of the best known of our British heroes is King Arthur and as he was a Welshman, it is appropriate that we should find several places associated with him in Wales, although the story is also connected with some English caves.

Dinas Rock at Pontneddfechan in South Wales is one of the sites of the sleeping hero story. This enormous triangular-shaped rock, a favourite of rock climbers and a practice ground of the Mountain Rescue, separates the Rivers

Sychryd and Mellte. Beneath the rock, there is said to be a cave where Arthur and his knights are sleeping in readiness for the fateful day. The story is that a wizard, meeting a drover carrying a hazel staff, asked him where the hazel tree grew. Digging up this tree, they found the cave which not only contained Arthur and his knights, but much gold and silver. The wizard told the drover that he could take as much gold and silver as he liked, but if he should touch the bell hanging by the entrance, the sleepers would wake. Should this happen, he must tell them to sleep on. Inevitably, the drover touched the bell twice and remembered the wizard's instructions, but the third time he forgot the words and the knights attacked him. He escaped, permanently crippled, but never found the cave again. A cave does, in fact, exist below Dinas Rock—a stream passage between the two rivers—which has been thoroughly explored, but without any sign of the sleeping heroes.

It is quite common for caves, castles and heroes to be linked together in legend, in the often mistaken belief that the hero was associated with the castle during his lifetime. Under Chepstow Castle, Gwent, a potter named Thompson was supposed to have found a cave in which Arthur and his knights were sleeping. A bugle and sword, the usual legendary items, hung on the wall, but when trying to blow the bugle and draw the sword, Thompson disturbed the sleepers and fled from the cave to the sound of a voice saying:

> "Potter, Potter Thompson,
> Hadst thou drawn the sword or blown the horn
> You would have been the luckiest man ever to be born."

Snowdonia has its Arthurian legend, for it is said that after Arthur's death, his knights went into a large cave called Llanciau Eryri on the slopes of Snowdon, adjoining Llyn Llydaw, to become sleeping heroes until Arthur's return. Arthur and his knights are also said to be sleeping under Richmond Castle, Yorkshire. Here the story is identical with that of Chepstow, even to the name of Potter Thompson. He also sleeps under Sewingshields Castle, Northumberland, and Mynydd Mawr (the big mountain) at Llandeilo, and Somerset claims that he is still sleeping at Avalon in the Glastonbury area. There is a small cave in the Wye Valley, (King) Arthur's Cave, but as far as I know this cave does not carry that particular legend.

A similar legend is told in connection with the red sandstone escarpment of Alderley Edge, Cheshire, although not a cave area. A farmer from Mobberley had a white horse for sale and was approached by a stranger, who took him to a cave. There, in the dimness of the cave, the farmer saw sleeping warriors, all with white horses, except one, who had no horse. The stranger explained that the farmer's horse was needed to make good this deficiency and the farmer might take a reasonable payment from the treasure in an inner cave. This he did and was told that the warriors would one day awaken to defend their country. He emerged into the daylight and turned to look back at the cave entrance, but it was no longer there. The legend associates the warriors with Arthur's knights and the stranger with the magician Merlin, whose name is attached to several caves in Britain, including the sea cave under Tintagel Castle in Cornwall.

Sir Walter Scott, in his *Letters on Demonology and Witchcraft* (1830), describes a similar legend in which the sleeping hero was Thomas of Ercildowne. A "daring horse-jockey" arranged to sell a black horse to "a man of venerable and antique appearance", who suggested that they completed the purchase at midnight on Luckenhare hillock in the Eildon Hills, Scotland. There the customer paid the horse dealer "in ancient coin" and invited him to view his home, a cave. Inside the cave were rows of stalls with a motionless horse and warrior in each. The horse dealer was told that these men would wake up for the Battle of Sheriffmuir, but on the wall was a sword and horn which had the power of dissolving the spell. The horse dealer tried to blow the horn. The horses stamped, the men awoke and a whirlwind swept the horse dealer from the cave which he never found again.

The story is repeated with the same location and some variations. One version concerns a horse dealer, Canonbie Dick, who had two horses for sale. He was paid on a series of occasions in ancient coin and finally suggested that the last payment be made at the stranger's home. The stranger agreed, warning him that if he lost his nerve he would die. He was taken to a cave where a horn and sword hung on the wall and was offered the choice of these. He could be King of Britain if he chose aright. Finally, he chose the horn, but when he tried to sound it, a hurricane carried him out of the cave and a voice told him that he should not have summoned help but have drawn the sword. The horse dealer was so badly hurt that he had only time to tell his story before he died.

Owain Llawgoch (Owen the Red-Handed) is said to sleep in a cave in the cliffs below Carreg Cennen Castle in Dyfed, Wales. The cliffs contain several small caves, the habitat of badgers, and some are difficult of access. Owain Glyndwr, the Welsh hero who in the fifteenth century made a last and unsuccessful attempt to gain Wales her freedom, is reputed to be sleeping in a cave in the Vale of Gwent.

Among the other sleeping heroes are Bruce, who sleeps in Bruce's Cave near Kirkpatrick Flemming in Dumfriesshire, although it is less than four metres in diameter, and Earl Gerald of Mullaghmast, who sleeps with his soldiers in a cave under Mullaghmast Castle in Kildare.

Usually, the time of awakening of the sleeping hero is indefinite, but sometimes the spell is for a certain period, as in the Devil's Hole, Isle of Man, where the slumbers of a prince will last six hundred years and one story, already told, has actually a date, that of the Battle of Sheriffmuir.

A common cave legend concerns a musician who enters a cave and his music is heard far underground until it fades away in the distance. The musician, most often a piper, is never seen again. In Carlingheugh Bay, Arbroath, Scotland, is the Forbidden Cave, where a drunken piper was lost and seen no more. In Galloway is Piper's Cave, and the music of the pipes was said to have been heard underground all the way from here to Barnbarrach, a remarkable feat as the cave, which contains a well some twenty feet deep, is less than forty metres long.

The Cave of Gold on the Isle of Skye has the identical legend and we find it again at Piper's Cave, Campbeltown, Argyll, where the piper was lost, but his dog escaped from the system at Southend, nine miles away. Some cavers explored it in 1945 and found that it consisted of eight small interconnecting chambers, certainly not so extensive as the legend implies, but perhaps the most persistent feature of these stories is the vastly exaggerated length of the system. One suggestion of the origin of the piper story is that the name is a misspelling of the Scottish word 'pipar', a priest, a name which dates from the time when caves were commonly used by early Scottish preachers. This theory would explain the frequency of pipers' caves in Scotland.

Sometimes, the musician is not a piper. In the Swansea Valley in South Wales, the lost man was an animal castrator who always announced he was in the neighbourhood by sounding a horn. He entered the Dark Well, the resurgence of Ogof Ffynnon Ddu, sounding his horn as he went. For some time the sound of the horn could be heard beneath the surface, but after it died away, he was never seen again. A way was forced into Ogof Ffynnon Ddu in 1946 and it was thought that nobody had ever previously entered it, but the day after its discovery I was asked to investigate a skeleton of a young man which the explorers had found coiled on a rock in one of the chambers. The right leg was bent at the knee and the other slightly extended, as if he had fallen asleep from exhaustion and died there. We looked round the chamber and concluded that the only way he could have entered it was through one of the two natural holes in the ceiling, through which he must have fallen in the dark from the passage above. There were no footholds in the eight-feet-high walls and it would have been impossible for him to have climbed out without assistance. Whether it was the animal castrator, we shall never know, for we found no remains of the horn and not even a metal eyelet or button to show the type of clothes he wore. Perhaps he was an early treasure seeker.

A drummer is still said to be heard under Richmond Castle, Yorkshire.

In a cave near Llanymynech, North Wales, the musician was a harpist. The cave or subterranean passage was supposed to connect the Lion Inn at Llanymynech with the church, and the harpist laid a wager than when the choir sang on the Sunday, they would hear his harp, although he would not be in church. At the appointed time, the sound of the harp was heard from the floor of the church. The harpist had made the journey underground from the cave entrance, but, as he was never seen again, he did not claim his wager.

It is possible that the beginnings of such stories may lie in speculation about the length and direction of the mysterious cave by local people who were afraid to explore it. The musician rarely came out again, so their fears were obviously well founded.

It is not surprising that caves were sometimes thought to house fairies and other supernatural beings. Fairy communities are often thought to dwell underground in prehistoric burial mounds, sometimes known as fairy mounds, such as the now destroyed New Stone Age burial mound at Nempnett Thrubwell, Somerset, which had the delightful name of Fairy Toot. There are many Irish stories of fairies and fairy mounds. The fairies lived underground, and from burial mounds to caves is an easy step. Many caves are known as the Fairy Cave, although sometimes this is in the context of a 'fairy grotto', meaning a cave with particularly pretty and delicate stalactites.

The cave at Llanymynech was linked with the fairy folk, who were said to wash their fairy linen in the water which ran through it.

At Rosehall, Sutherland, a man was enticed into a cave by music, which he heard coming from it. He was found in the cave a year later, dancing to the fairy music, and it was not until he emerged into the daylight did he realize that he had been away for more than a few minutes.

Not all fairies were mischievous, for at Runswick Bay, Yorkshire, a cave known as Hob Hole was supposed to be the home of a fairy who would cure whooping cough for the asking, but some were bad fairies. Boggart's Roaring Hole on Ingleborough, Yorkshire, was once thought to be occupied by a boggart, a name given in some parts of the country to a brownie who is mischievous or violent. A stone thrown into the hole caused the boggart to roar. Pwll y Rhyd in Wales also produces a similar noise, due to water disturbance, but has never been associated with a boggart to my knowledge. Hurtle Pot, Chapel le Dale, is thought to owe its name to a boggart, who causes people to hurtle themselves to death in the pot.

Some people have seen cave entrances as openings to the other world, the world of the dead, the Tir na NOg of the Irish or the Annwfyn of the Welsh, others as the route to Hell, the Devil and his demons, and yet others as the entrance to fairyland. It was said that there was once such a cave entrance in Cwm Llwch, on the northern flank of the Brecon Beacons. This was the entrance to fairyland and on one day in the year, the fairies held an "open day" when mortals might visit them. All went well for many a year, until one human vandal picked a fairy flower, which so enraged the little people that they closed the entrance to fairyland and nobody has ever been able to find it again.

There were once said to be fairies in a cave at Chudleigh, Devon, and their activities can be best summed up by the sonnet on the cave, written by the Rev. John Swete in 1794:

> With this vaulted rock, the Fairies dwell
> I fear to tread upon the enchanted ground
> For o'er this arch and through the Cave's dark round
> They've spread a charm; and wove a mystic spell.
> Ev'n now perchance they scoop the concave roof,
> Or pave with sparry studs the fretted floor
> Or hang their icicles of stone aloft
> Beneath yon branchy oak, on moonshine nights
> They thread the dances or in yon streamlet lave,
> Within the precincts of the hallowed Cave
> They celebrate alone their festal rites;
> Ah then, rash Mortals dread too near to press
> For vengeful Phantoms guard the deep recess.

British caves seem remarkably free from 'monster' legends, but there is one about monstrous eagles connected with the Red Castle, Castell Goch, near Taff's Wells just outside Cardiff. The story goes that there is a cave and tunnel which connects the castle with Cardiff Castle. The cavern is reputed to contain an iron treasure chest which belonged to Ivor the Little. This chest is guarded by three great eagles and when people attempted to reach the chest in the seventeenth and eighteenth centuries, they were beaten off by the huge birds.

Thief's House, a cave at Fitful Head on Mainland in the Shetlands, was the home of Black Eric, a cattle or sheep thief who kept a kelpie—a sea-horse—named Tangie. Black Eric had a fight with a crofter Sandy Breamer who was about to overcome Black Eric, when Tangie picked up both combatants and swung them about in the air, but in spite of his giddiness, Sandy managed to kill Black Eric by knocking him from Fitful Head. Tangie the terror sea-horse continued to haunt the area.

Dogs are more common animals in cave stories. The story of the piper and his dog at Campbeltown, Argyll, seems to be a combination of two legends, the piper story and the dog story. The account of a dog entering a cave and reappearing miles away, sometimes with hair standing on end or having lost it altogether by squeezing through narrow passages, is a common tale told by many guides to impress visitors to show caves. Underground systems are often credited with being miles long, but when explored their extent is disappointingly modest or in some instances they do not exist at all.

The dog legend is associated with Hoyle's Mouth, West Wales, which P.H. Gosse in his *Tenby: a seaside holiday* of 1856 stated was thought to connect with a cave called The Hogan under Pembroke Castle eight miles away and that "the vulgar" are very averse to exploring even its mouth "although a gentleman is said to have penetrated a considerable distance and found fine rooms". Many others, including myself, have followed in his footsteps, or perhaps I should say the hands and knees of that gentleman, and it would be wrong to say that one can go any distance. The cave contains an interesting small terminal chamber and connecting passage, but nothing so impressive as the entrance. A small passage continues from the terminal chamber, but although a small dog might squeeze in, it is certainly too small for a caver.

The Jug Hole caves of Matlock, Derbyshire, have been associated with a dog, or rather the sound of one, for when miners broke into the natural cavities over a century ago, they were so scared by what sounded like the angry barking of a dog that they withdrew. In the 1940s a climbing party decided to investigate the noise. Part way along one of the cave passages, they heard the barking ahead, which stopped after a time. The passage ended in a small chamber with a pool through which they waded and then sat down on the other side. What had appeared to be a quiet pool suddenly became agitated and after a series of loud booms ending with a roar, the water disappeared down a hole. The stream in the cave began slowly to fill the pool again and when it reached a definite height, the disturbance began all over again. The noise caused by the displacement of the water sounded in the passage outside the chamber like the barking of a dog.

Distant noises in a cave, because of the acoustics, are often warped beyond recognition. The voice of a guide in Speedwell Cavern in Derbyshire, talking to his passengers as their boat slides along the narrow flooded mine passage, becomes unintelligible to anyone at the other end of the tunnel and the sound heard becomes a weird inhuman muttering.

The phenomenon of the Jug Hole is identical with that of Wookey Hole, Somerset, where the noises are heard at very infrequent intervals and years elapse between the incidences. Although I was the archaeologist at Wookey Hole for over twenty years and spent many hours of the day and night there during excavations and diving operations, I never heard them. My predecessor, Mr H.E. Balch, heard them on several occasions. His first experience was on an occasion before the First World War. He was standing with his dog and a party of people near the Witch stalagmite, when they heard threshing noises coming from the rising water beneath the arch near the Witch. As the dog had left the party, they thought it had got into the flooded tunnel, but when the noise stopped, the dog appeared, quite dry, from another direction. His next experience of the noises was in 1910 when with a friend in the second chamber, the 'Hall of Wookey', they thought they heard a party of people coming into the cave, but as the sound became louder it sounded like a chariot coming towards them. The noise grew so overpowering that they were about to run when it suddenly stopped.

On 6 September 1911, Harry Savory, an old friend of mine who did much of the photography of Wookey Hole and whose photograph of the Witch is known far and wide, was in the cave with his brother. Like Mr Balch, they were standing by the Witch when they heard a faint, high-pitched hammering from behind the walls of the cave. The hammering grew louder, lower in pitch, but not so rapid and then seemed to die away and stop near them. The sound apparently did not reach a frightening stage as on some other occasions and was similar to the water-hammer which sometimes occurs in domestic water pipes.

In 1913, the guide, who had conducted many parties through the cave, heard loud hammering coming from the Third Chamber or Witch's Parlour, but did not investigate. In 1913 too, Mr Balch and Mr Savory were in the cave with a party photographing formations when they heard a throbbing but musical sound coming from the same chamber. They rushed into the Witch's Parlour, but were only able to hear the faint remnants of the sound, as they died away somewhere in the then unknown passages beyond the Parlour. They suspected correctly that the noise was linked with the rise and fall of the Axe which flows through the cave. A sluice gate has existed for many years at the resurgence of the cave and when the local paper mill was working, it was usual to open the sluice during the lunch hour. As the noises corresponded with that time, the party decided to see what would happen an hour later. Sure enough, the sounds occurred again, increasing in intensity and dying away, but as they were then in the Third Chamber, they were able to distinguish another sound, that of hammering, with a low roar or rumbling in the background. Later in the day, when in the Third Chamber, they heard the sounds again, but they also noticed that rings of ripples were forming on the surface of the water.

Wing Commander Hodgkinson, a former owner of the cave, had a similar experience in the cave when he heard a plopping noise from the Third Chamber. He found the Chamber flooding as it does on occasions and realized that the noise originated from large bubbles forming and bursting on the surface of the water. These were formed by air displaced from waterlogged passages and the noise was exaggerated by cave acoustics.

It was possible to create the noises at Wookey Hole when the water was at the right height, as it occurred during a diving operation in the resurgence when the displacement was sufficient to cause an escape of air within the system, and although the noises were not heard by the divers, they were clearly audible to people within the cave. The sluice gate was no longer used as the mill had ceased to operate, but although this was no doubt a contributory factor on some occasions, we know that the noises were heard long before the sluice gate was installed, for even Clement of Alexandria, in his *Stromata* of the late second or early third century A.D., wrote of a cave in the side of a mountain in Britain that when the wind blows into the cave, it causes a noise like the clashing of numerous cymbals. Although the wind produces noises in Fingal's Cave in the Island of Staffa, his description of the cave leads us to believe he was writing of Wookey Hole.

The Great Cave of Wookey Hole is known throughout Britain for the legend of the Witch of Wookey. The Witch is the name now given to the huge stalagmite boss, overlooking the River Axe in the first main chamber, known as the Witch's Kitchen. The likeness of the witch can be seen from two directions, but the best-known view is the profile, which has some resemblance to a head and shoulders. The head has a particularly prominent nose and is best seen in silhouette with a light behind it. The legend is that an old recluse once lived in the cave and was credited with witchcraft. She is said to have cursed the people in the neighbourhood, causing their cattle to die. The source of the River Axe lies in an old lead-mining area and was said to have been polluted at one time by waste water from ore washing. It is possible that sickness in cattle may have been due to lead deposits in the water, particularly as the first reference to the stalagmite figure as a witch appears to have been in 1694, during a period of lead-mining activities. William of Worcester in 1470 described a "figure of a woman—clad and holding in her girdle a distaff". Of course the witch was also blamed for other calamities, such as the lack of potential husbands, according to Dr Harrington of Bath in his poem published in 1756, which tells the story of the legend. These extracts describe how the witch was eventually turned into stone:

> Deep in the dreary dismal cell
> Which seemed and was ycleped hell,
> This blear-eyed hag did hide:
> Nine wicked elves, as legends sayne,

She chose to form her guardian trayne,
And kennel near her side.

From Glaston came a lernede wight,
Full bent to marr her fell despight,
And well he did, I ween:
Sich mischief never had been known,
And, since his mickle lerninge shown
Sich mischief n'er has been

He chauntede out his godlie booke,
He crost the water, blest the brooke,
Then—pater noster done,
The ghastly hag he sprinkled o'er:
When lo! where stood a hag before
Now stood a ghastly stone.

But tho' this lernede clerke did well:
With grieved heart, alas! I tell
She left this curse behind:
That Wokey nymphs forsaken quite,
Tho' sense and beauty both unite,
Should find no leman kind.

Mr H.E. Balch who carried out the major excavations at Wookey Hole discovered the remains of a human skeleton, the bones of two kids and a milking pot, just within the entrance to the Great Cave. These he thought were the relics of an old recluse, the original witch of the legend. However other skeletons belonging to Roman times have been found in the same locality, together with sherds of Roman pottery and pre-Roman ware, similar to the pot. If the bones of the 'recluse' belong to the same context, they would be of a much earlier date than the witch legend.

When human bones were found on the river bed during diving operations, it was suggested that they might have been the remains of sacrifices to the witch at some gruesome ceremony performed before the witch stalagmite. The skeletons from the river were, however, of Romano–British origin and found both down and upstream of the Witch, so could not have all been washed downstream from the stalagmite. The source of the bones was traced upstream to Chamber 4, the Holie Hole, where one upper arm bone of a woman and part of the shoulder, together with two beads of a necklace, were found *in situ,* and other relics subsequently. This chamber was clearly a burial place during the Romano–British occupation and was being eroded by the Axe.

There was probably a poor old woman, living at some time in the Great Cave, whose appearance, solitary habits and strange dwelling led the local villagers to look upon her as a witch. There are several caves where the outcast, the misfit or penniless have lived in caves. Hagtorn Cave, Whernside, is said to have been torn open by a witch.

Of all the categories of cave legends, perhaps the 'oracle' story is the oldest. This has a very obvious connection with classical mythology. The cave is occupied by an immortal, a solitary fairy, who can be consulted on love and other matters. These stories may well survive from the old pagan belief in the cave god or goddess or water nymph presiding over the stream issuing from the cave.

Such a cave is Tresilian Cave in Tresilian Bay, Llantwit Major, Glamorgan, said to be occupied by an immortal named Drynwen, whom lovers could consult if they could throw a stone over the natural bridge in the cave without hitting the ceiling. Unfortunately the formula which had to be recited is no longer known. Even if you knew the formula, you would have to be there at the right time, as at high tide the sea can cover the bridge which is about seven feet below the roof. It is said that the cave connects with St Donat's Castle, a short distance to the west. Such underground passages are common legends. About 1760, Mr Picton of Poyston was married in the cave to Miss Powell of Llandow. They became the parents of General Picton. The Bishop of Llandaff suspended the minister who was responsible for the ceremony and the couple were remarried in the more traditional way at Llandow. At the time of the ceremony, the cave was known as St Julien's Cave, but was nicknamed 'Reynard's Church', perhaps an allusion to the priest who carried out the marriage ceremony there. Today it is known as Tresilian Cave.

Caves are often thought to have treasure hidden in them and before banks existed they were as good hiding places as any. A number of pots containing Roman coins have been found buried in fields and so have many gold and silver articles. Probably the owners died without telling anybody where they had hidden their wealth and a local cave was given the credit of being the hiding place. Many coins and objects have been found during cave excavations, but I do not know of any find which could really be called treasure. Many of the treasures reputed to be in caves are as fabulous as that of Aladdin, although the finding of them never does anyone any good. They turn out to be illusions and the caves are never found again.

There is a legend of this type of cave near the Nant Ffrancan Pass near Bethesda in North Wales. In the dim past, a man found a cave at dusk and, as it was rumoured that there was a cave in the district containing a vast treasure, he decided to explore the cave the next day. To find the entrance again, he was careful to lay a trial by cutting splinters from his stick, but during the night, every piece was taken away by the fairies and the cave was never found again.

Pirates are often thought to have buried their treasures in some sea caves, but the chances are that they probably made good use of their ill-gotten gains during their lifetime. There is a story that Captain Kidd hid a treasure amounting to £300,000 in a cave on Steep Holme, a lonely island in the Bristol Channel. As I was a trustee of the island for over twenty years, I know every cave on Steep Holme. They are either sea-washed or do not contain enough soil to cover the smallest treasures. Kidd's activities as a pirate were in more distant parts and it is likely that the only time the Scottish-born sea rover ever came to Britain as a pirate was when he was brought from America to London to stand trial and then be executed. He presented part of his booty to the Governor when landing in Boston, just before his arrest, apparently in the hope of avoiding the consequences of his crime, which would indicate that he carried a substantial part, if not all, of his treasure with him.

Another category of cave legends deals with giants and the Devil which have one thing in common, the size of their exploits. The giant Fingal occupied a cave on the Island of Staffa and there he built a road from the cave to the Giant's Causeway in Ireland. The rock at Fingal's Cave and the

Giant's Causeway is identical basalt, connected geologically, and this is an example of a legend growing up to explain a natural phenomenon.

When floodwater issues from the Peak Cavern, Derby-shire, legend attributes this to the Devil relieving himself.

Leaving legends aside, we now come to associations with caves which can barely be called historical. We do not know whether the events actually took place, but they do concern real people, princes, highwaymen, outlaws, soldiers, saints, murderers and just poor cave dwellers.

Caves have always formed convenient hiding places for fugitives from enemies or the law. Llywelyn the Last, the one and only official native Prince of Wales, was forced to spend a night in a cave in the Aberedw Rocks near Builth, when hunted by the King's troops during the last days of his life. There is a story that when he was at Aberedw, he bribed a smith to reverse his horse's shoes so that his pursuers would be misled by the hoof prints in the snow, but his enemies forced the smith to betray his prince. This impracticable ruse is not only told of Llywelyn, but also appears in romantic stories of fleeing highwaymen.

There are highwaymen's caves, for they were, no doubt, ideal hideouts for such rogues. When a boy, I used to play in a shallow, artificial chalk cave on Shooter's Hill in south-east London. It was known as Dick Turpin's Cave, the highway-man who never rode his horse to York. I doubt whether in fact he ever used the cave, as an adjoining kiln would suggest an agricultural lime working. Poole's Cavern, now reopened as a show cave at Buxton, was supposed to have been the retreat of a highwayman of that name.

Another outlaw, perhaps better described as an outlaw hero, was Rob Roy, or Rob the Red, because of his red hair, to use the Gaelic meaning of his name. His real name was Robert McGregor, although he was forced for the time to use his mother's clan name of Campbell, when his own clan of which he became the head in 1693 was outlawed. In 1712, the Duke of Montrose seized his lands in payment of debt and Rob Roy and his clansmen retaliated by harassing the Duke and stealing his cattle. Rob Roy was said to have robbed the rich to pay the poor, like the English Robin Hood,

whose legendary origins go back much further than the birth
of Rob Roy. It was inevitable that Rob Roy should, at some
time, have found a hiding place in a cave, but whether his
choice was the cave which bears his name in the Drymen/
Aberfoyle district of Scotland is a matter of conjecture.

W.ales also has its Rob Roy, Twm Siôn Catti, and the cave
named after him is near Ystrad-ffin, Dyfed. He was a
sixteenth-century landowner turned outlaw, who robbed the
rich to give to the poor. Once he is said to have deceived
another highwayman by pretending to be a farmer, whose
tired horse carried bulky saddle bags filled with shells. The
highwayman, believing they must contain gold, challenged
the 'farmer', who with feigned alarm threw the bags over a
hedge. While the highwayman was retrieving them, Twm
Siôn Catti mounted the highwayman's horse and rode off
with saddlebags filled with loot.

Caves were valuable hideouts during civil wars and local
skirmishes. Bonnie Prince Charlie took refuge in more than
one cave in Scotland.

John Cann, a royalist, hid from the roundheads of Bovey
Tracey in Devon in a cave known as Bottor Rock Cave until
he was routed out by bloodhounds and taken to Exeter for
execution. He was said to have left treasure in the cave, but
like most treasures, it has never materialized. John Cann still
haunts the district by night.

A Lancastrian, Dafydd ap Shenkin (David, son of Jenkin),
was said to have hidden in a cave near Dolwyddelan Castle, a
traditional birthplace of Llywelyn the Great, Llanrwst,
Snowdonia. The cave was later called Ogo Dafydd ap
Shenkin, as was Lord Huntly's Cave near Grantown,
Scotland, named after a Marquis of Huntly who hid in the
cave after fighting with Charles I's troops against the
Covenanters.

Squire Elford of Devon, whose family occupied a mansion
adjoining the site of the later Plymouth Reservoir, hid from
parliamentary troops in a very small cave near the top of the
nearby Sheeps' Tor. In Ireland, Lugan, Earl of Desmond,
took refuge in the old Mitchelstown Caves, Co. Tipperary, in
1601, but he was betrayed by a relative in return for the
£1,000 reward. Lugan died in the Tower of London. It is said

that his foster-brother was killed during the skirmish and his body left in a wooden coffin in the cave. It was reported to be still there as late as 1908.

Besides their use as temporary refuges, caves have been used as permanent homes throughout the ages. Ready-made shelter, with an almost even temperature away from the entrance and the security of only one way in, no wonder caves were much-sought-after residences in ancient times. They provided suitable winter lodgings for the hunting families of the Old Stone Age, who left ample evidence in the form of flint implements, hearths and food bones. Even when people had learned to build, many still lived in these natural homes and the New Stone Age agriculturists, Bronze Age and Iron Age peoples have left their rubbish and their bones. Even in recent years and indeed sometimes today, caves are still used as dwellings. The Spanish gypsies of Sacro Monte, Granada, adopted artificial caves and there are still some troglodyte families who live in holes made in the surface of the Tunisian Desert.

A family named Evans lived for a number of years in a cave now known as Pride Evan's Hole in Cheddar Gorge. The two Glynn sisters occupied a cave at Port Patrick, near Cemaes, Anglesey, which was to become known as Jenny Glynn's cave or Ogof Sian Glynne. Sian (Jane) was the elder sister. They made a living by travelling the countryside selling Port Patrick white clay, which was then in demand. In East Mendip, there is a Nancy Camel Hole, although there seems to be little known about Nancy Camel. Many caves bear the names of their once occupiers, but their stories have been forgotten.

Perhaps one of the best-known names is Mother Shipton, who was born in a cave at Knaresborough, Yorkshire. Her maiden name was Ursula Southell and she was brought up an orphan in Knaresborough to become a famous prophetess of the sixteenth century. Even as a child she seemed to be surrounded by supernatural forces and, in spite of her ill-looks, she married Tony Shipton in 1512. She was famous for her prophecies, although many of them were of late nineteenth-century invention.

The Cave of the Blue Horse near Penryhn, Cemaes Bay,

Anglesey, owes its name to a tragedy which was supposed to have taken place in the seventeenth century, when John Wynne, the heir to an estate at Cromlech, having supported Charles I, fled to France. As it was thought that John had died there, his brother Robert inherited the estate and persuaded John's former fiancée, Margaret Lloyd, to marry him. Some years later, John returned and was so changed in appearance that nobody recognized him. He made himself known to his brother and his brother's wife and asked his brother to lend him a horse to help him leave the country. The next day, the horse was found dead in the cave and John's body was nearby.

Taylor's Hole, Bridgnorth, is supposed to be named after a tailor, who climbed up into the cave to make a coat there for a wager. He was just sewing the last button when the thimble fell down the cliff face. In trying to recover it, the tailor fell and was so injured that he was unable to finish his work and so claim his wager.

At Cove Haven, Arbroath, Mason's Cave is not named after one of my own ancestors but after St Thomas Lodge of the Freemasons, who used it occasionally as a meeting place.

A good deep pothole was a convenient place for a murder. Noon's Hole in the county of Enniskillen is a 300-foot-deep pothole and it is said that down this hole was thrown a man named Noon, murdered by smugglers because he informed the excisemen of their activities. At Knaresborough in Yorkshire, a cave was used by three robbers, Aram, Clark and Houseman, as a refuge and hiding place for their loot. Eugene Aram murdered Daniel Clark during a quarrel in 1745 and buried him in the cave. Fourteen years later, the crime was discovered and Eugene Aram was executed at York.

Some caves are reputed to have been occupied by hermits or saints, a legend which is world-wide. St Ninian's Cave, near Whithorn, Galloway, is named after St Ninian, who is said to have used it as an occasional retreat. The discovery of incised crosses on the rock and rock debris some years ago is thought to authenticate the story.

In the Swansea Valley, clearly visible from Dan yr Ogof Caves, across the valley of the Tawe, is an impressive opening, although not so large as it appears. It is a single chamber with an outward sloping floor and is known as Eglwys Caradog or Cradoc (the Church of St Caractacus). The name 'Caractacus' is deceiving, as it has no connection with the warrior chieftain of history, the son of Cunobelinus, the Cymbeline of Shakespeare. The legend of Eglwys Caradog concerns Cadicus or Cattwc, the son of Cynlais. Cynlais, who later became a saint, was the ruler of the valley and his name survives in the name of the town Ystradgynlais (the Vale of Cynlais) a little further down the valley. It is reputed that he died in his son's arms within this cave.

Caves with their 'bottomless' pits and endless dark passages must have always been sources of wonder and speculation to local people, and the great caves, such as Wookey Hole and Peak Cavern, also fascinated visitors from afar, but it was with the nineteenth-century's romantic interest in follies, ruins and grottoes that caves became fashionable. In the latter part of the eighteenth and the first half of the nineteenth centuries, it was the habit of topographical writers to roam the remote areas of Britain in search of the 'curiosities of nature' and of course caves came in for their fair share of attention. One writer in 1798 described Porth yr Ogof, Wales, as lying in 'a profound and gloomy glen'. The cave itself, according to various descriptions, was full of 'bottomless gulfs' and 'stygian horrors', and all was 'horribly grand'. 'Portal' was the word used for the cave entrance.

When Banwell Caves, Somerset, were excavated in the early nineteenth century, the area round them was landscaped in the romantic fashion and, not content with a natural cave, two small artificial grottoes were made nearby. Within the built-up stone recess of one of these is an oval stone table, while over the entrance a rectangular marble plaque is inscribed:

HERE WHERE ONCE DRUIDS TROD, IN TIME OF YORE
AND STAIN'D THEIR ALTARS WITH A VICTIM'S GORE
HERE, NOW, THE CHRISTIAN RANSOMED FROM ABOVE
ADORES A GOD OF MERCY AND OF LOVE.

SEVEN

Caves of Ingleborough and Stump Cross

WITH A HUGE AREA of limestone suitable for cave formation and consequently a large number of caves, Yorkshire has attracted cavers from earliest caving days. Because of the great number of potholes, i.e. deep shafts going down vertically from the surface, the northern caver is usually known as a 'potholer', while in the south, the cave explorer is usually described as a 'caver'. Occasionally the press refers to both northern and southern cave explorers as potholers, but technically this is incorrect. There are of course exceptions to this general rule and there are a few potholes to be found in the south and cave entrances in the north which are not pots. There are also internal shafts or 'pitches' in the south, but it is not possible to even begin to explore many systems in the north without first descending a pot.

Geologically, cave systems are known as 'karst', a German word given to the Yugoslavian cave area. This area is regarded as the typical area with which all other cave areas are compared and the formation of limestone caves is often described by the imposing term of 'karstic phenomena'. The Yugoslavian Karst is a bare rock area, fissured and containing a great many depressions known as 'dolines', indicating subterranean cave systems. Perhaps the nearest approach in this country to the Yugoslavian karstic surface features is that of the Yorkshire 'clints', although they cover a much smaller area. The Yorkshire clints are bare and fissured. The clints refer to the general surface appearance, while the fissures, some of which develop into potholes, are known as 'grikes'. The principal potholing area of Yorkshire is that enclosed by the heights of Whernside, Penyghent and Ingleborough.

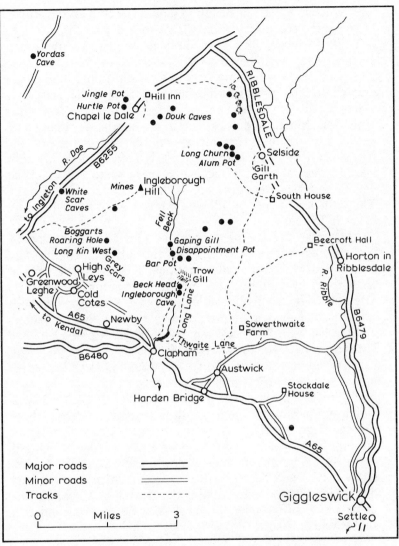

CAVES IN THE INGLEBOROUGH AREA, YORKSHIRE

The deepest shaft in Britain, Gaping Gill, defied the efforts of the early potholers to scale it. It took a Frenchman, E.A. Martel, to make the first descent of the 360-feet-deep entrance shaft on 1 August 1895. Martel's exploits in France and elsewhere earned him so much esteem that a statue was erected to him in his native town, probably the only case where a statue has been provided for a caver's speleological activities. Martel's descent fired an enormous amount of enthusiasm among the earlier cavers and, although the Yorkshire Ramblers were particularly active before Martel's visit, in the immediate successive years some of our oldest caving societies were formed, both in the north and south. The names of Gaping Gill and Martel are known to every caver in Britain.

Martel wrote a complete chapter on his descent in his book *Ireland and English Caverns*, for not only was Gaping Gill familiar to Martel, but also practically every cave known at that time in Ireland and England. Among other early attempts to scale this formidable pothole, Professor Hughes's efforts had ended in failure, and forty years before that in 1842, a Mr M.J. Birkbeck, a member of the Alpine Club, had got part way down the shaft when one of the strands of his rope frayed on the edge of a rock and he had to be hauled up again, somewhat wet, although a trench had been dug over half a mile from the lip of the pot to discharge the Fell Beck, which tumbles down Gaping Gill, into an alternative pothole, Grange Rigg Pot. Eight years later, Birkbeck made a second attempt and, although he was lowered to a ledge which still bears his name, about 190 feet down, the rope again became badly frayed and he was pulled up again. Although the beck had been diverted once again, the ascent of the shaft was made more difficult by the force of water pouring out of a subterranean fissure.

No further attempt appears to have been made until Martel arrived in 1895 and the reputation of the "terrible gouffre" was challenge enough to the ardent explorer. He wrote that to descend it was his greatest desire and the principal objective of his 1895 visit. Some months previously, Mr Farrer, the landowner, had promised his full co-operation with the project and when Martel arrived at Clapham, he

The Dream Cave, Treak Cliff Cavern

Ogof Pwll y Rhyd

The Cluster, Straw
Chamber, Dan yr
Ogof

The Flitch of Bacon,
a curtain at Dan yr
Ogof

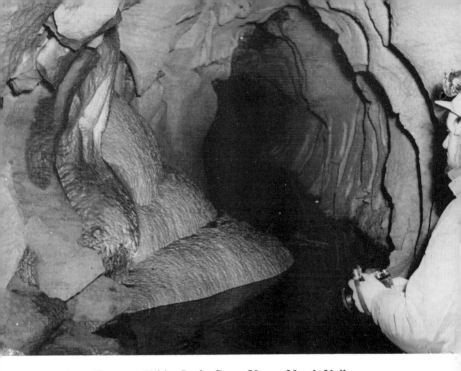

Gour Passage, White Lady Cave, Upper Neath Valley

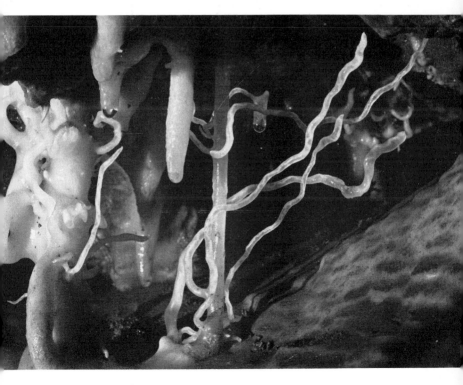

Helictites in a Mendip cave

Fixed ladders to Valentine Series, Lamb Leer, Mendip

Lower Stream Passage, August Hole, Mendip

Reed's Cave, Higher
Kiln Quarry,
Buckfastleigh, Devon

Gruffy ground,
Harptree, Mendip

Old tubs in Black Engine Mine, Eyam, Derbyshire

Lead resmelting flues, Charterhouse, Mendip

Magpie Sough, mine
drainage level,
Derbyshire

found that the work of reopening Birkbeck's old diversion trench was in hand, as the beck was in spate owing to heavy rain during the last week of July. Transport and a support party had been organized.

At 10 a.m. on 30 July, Martel, in company with others, inspected the entrance of the pot and, in view of the amount of water going into it, thought it would be impossible to make the attempt. As he himself stated, it looked as if his dream was "drowned". The weight of water was heavy, and it was possible it would suffocate him or knock him off the rope ladder and might also cause the ropes to fray against the rocks. However the weather began to brighten and the beck to subside rapidly. Mr Farrer suggested that the attempt be deferred for forty-eight hours so that he could complete the trench and so reduce the amount of water going over the ledge of the pot.

For the next two days, Martel visited other local caves and at 10.30 a.m. on 1 August, he returned to Gaping Gill, where about a hundred spectators had gathered. It was threatening to rain again. The trench did not divert the whole of Fell Beck, but only about a twentieth of the previous volume of water was pouring into the pot. What was going down sounded formidable, but not insurmountable.

The depth of the pothole had been plumbed and it was found that the rope ladder was 65 feet too short. As he had previously done on certain of the French caves, Martel decided to descend the first 65 feet on a rope and use the ladder on the lower part of the pot, so that the more difficult section of the climb would be nearer the top, where help would be at hand. Martel was impressed with the methodical way in which each of the support party carried out his duties and the silence of the spectators. He wrote afterwards of his appreciation of the "flegme britannique" and compared this event with some of his experiences in France. When he descended Mas-Raynal d'Aveyron, the local lads organized a country dance accompanied by accordion and violin on the edge of the pothole. There was not only the risk of the participants falling in, but the noise prevented anybody hearing the orders being given over the subterranean telephone. On another occasion at Padirac in 1890, the police had to be called to remove a hunter who insisted on approaching the edge and whose dogs were causing stones to fall down the shaft. As Martel said, "My public at Gaping Gill were certainly the quietest and most serious that I have ever come across."

It took two and a half hours to join up the ladder sections on the surface, prepare the telephone, explain the procedure to the support party and prepare and fix supports for the ladders. At one o'clock, held by a rope round his waist, Martel, with his arms full of rope ladder, climbed down to the rocky edge of the shaft and threw the ladder down the pit. At the bottom end another rope had been attached should the ladder still prove too short. For lighting Martel carried a lantern suspended on his arm, while, tied up tightly in oil cloth as a protection against water, he carried candles, matches and magnesium. He was also equipped with a container of rum.

At 1.22 p.m. with his wife installed at the telephone at the top of the shaft to relay Martel's orders to the surface party, he put on his leather helmet and started to descend. For the first 65 feet, the descent went well. The falling water was five feet to his left and the only damping effect was the spray. He reached the first rung of the ladder without pausing and then started to climb down through the waterfall. The water went down his collar and back and he wished he had holes in his boots to let it out. He found the lantern and telephone rather cumbersome.

About 130 feet down, progress was stopped as a knot joining two ropes together was caught in a rock crevice and he had to remain stationary in the water chute until it was dislodged. At 190 feet down he reached Birkbeck's Ledge, where he was able to shelter from the water while he disentangled the lower half of the rope ladder, which had fallen on the ledge when thrown down the pothole.

The noise of the falling ladder and the stones it dislodged was heard by the spectators above, who thought that something terrible had happened below, but they were reassured by the surface party who were in communication with Martel and had secured the lifeline on his instruction. Requesting the surface party to let the lifeline down slowly and evenly, Martel continued his journey down through the waterfall. At last, 250 feet down, the shaft opened up and he found himself swinging in an immense void. He had reached the ceiling of the huge chamber below the shaft. As happens with a free-hanging rope ladder, it twisted and swung as he descended it, taking its climber in and out of the waterfall. He let his lantern fall as it was cumbersome and he had no further use for it. Suddenly there was another stop. The lifeline was found to be too short and the surface party announced that they would tie on another. Once again, the enforced stop left Martel dangling under the full force of the water. At last the attachment was made and he descended to the floor of black sand and pebbles. So close had been the previous sounding, that the last rung of the ladder was only 8 inches from the floor, but as soon as he took his weight from the ladder, the gap increased to about three feet. Mr Farrer, who had been keeping a time check, announced that the descent had taken twenty-three minutes.

Martel was the first person to stand in the vast main chamber, visited by many hundreds of cavers since that time, although descents are now often made by winch, fitted with an arm to swing the bosun's chair away from Birkbeck's Ledge.

On the floor of the chamber, Martel announced by telephone that the surface party could have their lunch as he was abandoning the telephone for at least half an hour while he explored. In the chamber, Martel summed up the characteristics of the pothole. It was similar to those to which he was accustomed in the French Causses, except that they were inactive, while the water action at Gaping Gill was still very much alive. There were two waterfalls, the one which descended vertically from the beck above and fell uninterrupted to the floor and another, which caused Mr Birkbeck such inconvenience, issuing from a fissure in the shaft. Fortunately for Martel, it was not running at the time of his descent, because of the work carried out by Mr Farrer. Martel described the chamber as large enough for a cathedral if the spire was pushed up the shaft. Indeed, he was much impressed with this chamber, which is about 475 feet long, 81 feet wide and 150 feet high. He likened the waterfall as seen from below to an oval column, like a moving stalagmite, and the prismatic effect of the daylight filtering from above through the water was one of the most extraordinary sights he had ever seen. He arranged the candles he had brought with him at various points to help assess the shape of the chamber.

He would have attempted a minor exploration of a passage leading off to the south-east where he could hear the sound of a stream, but he dare not do so without a companion, as a slip on the boulders might prove fatal in such conditions. At the foot of the talus at the entrance to the passage, Martel picked up a food tin, but many other objects, known to have been thrown down the shaft from time to time, must have been completely obliterated by the gravel and pebbles. In the north-west angle of the chamber was a yard-wide opening which could be penetrated for about sixty feet and which he thought might continue if the blocking sand was removed. He had hoped to find an alternative route out of the cave, but found none. After an hour and a quarter Martel, wet from heels to ears, as he described himself, teeth chattering and rum flask almost empty, picked up the phone for the party above to draw in the lifeline as he ascended, but the phone was full of water. He kept on shouting "Pull, pull!" At last he was heard and suddenly he was pulled up so fast that he could hardly get his hands on the ladder. Then came a sudden stop as the knot of the extension piece of the lifeline jammed in a crevice. Once again he was in the cascade, but this time he was much colder than before. At first he tried to concentrate on the scene around him, but the cold became unendurable. He tried to climb the ladder without the lifeline, but his legs were stiff and he was unable to climb more than four or five rungs, but in doing so he loosened the tension on the lifeline, enabling the knot to be disengaged from the crevice. Then the lifeline was drawn up again at quite a rate, but when he reached Birkbeck's Ledge, he was able to call to the party above to stop pulling.

For the rest of the ascent, communication was by voice, which was just as well, because no sooner had the surface party started to pull up the lifeline again, than the telephone cable became entangled between a projecting rock and the ladder and broke. At about 30 feet there was a further stop in the waterfall while the last knot was disengaged, and then after twenty-eight minutes for the full ascent, Martel was on the surface. His wife told him that she had in fact heard all his remarks through the telephone up to the time the cable broke, since the only fault was in the flooded receiver.

When they came to draw up the ladders, it was found that the tether had become securely jammed in a crack and human effort could not move it. A rope was attached to the tether and fastened to two horses used with the cart for carrying the gear. The horses not only dislodged the tether but pulled the whole of the ladder and ropes out of the shaft. Twenty minutes after Martel emerged from Gaping Gill, a storm broke and he was thankful that he was not at the bottom of the pothole.

E. Calvert, a member of the Yorkshire Ramblers' Club, founded in 1892, had intended to make the descent and spent nearly a year before Martel's descent considering the project. Martel regretted that he did not know about this, as he would have proposed a joint effort. However, although Calvert was unable to share the credit of the first descent of Gaping Gill, he was able to make the second descent. In the month following Martel's success, he began a series of assaults on the pothole. The first attempt in September 1895 was a failure. He spent some time building a dam, a method probably not so reliable as a diversion, because of the risk of collapse and so releasing large quantities of water far in excess of the normal flow. Calvert and his friend Booth then descended the shaft by bosun's chair, controlled by a windlass. Booth descended as far as Birkbeck's Ledge, but Calvert managed to get to 210 feet when the ropes became entangled and he had to be drawn up again. The remainder of the party consisted of nine people, Barran, Bellhouse, Booth, Cuttriss, Grey, Green, Lund, Thompson and Mason. There was no time that day to make a further attempt.

Calvert's next serious attempt was not until 9 May 1896. He had intended to make a further attempt the previous year, but the beck was too full and instead they examined the pot with a view to finding an easier method of descent. Close to the mouth of the pothole is an opening to what was formerly called Jib Tunnel Passage, which comes out in the main shaft of Gaping Gill. Martel had been unable to get his ladders into this tight passage but Calvert threaded a series of ropes and pulleys through the passage to the main shaft and he and his friends descended to the bottom by an elaborate arrangement of rope and pulley. Fortunately there was little water. otherwise they may have had the same trouble as I myself had when we tried a similar method at Swildon's Hole, Somerset, where the whole equipment jammed when a man was halfway down the shaft because the ropes were swollen in the pulleys. Calvert's only mishap was to fall into a pool of water which put out his light. As time was getting on, he did not stay to explore, but ascended immediately, making the return journey by winch in four to five minutes.

On the following day, Calvert made the descent again, with Booth, Cuttriss, Grey and Green, four of the people who had made the first attempt with him in September 1895 and all members of the Yorkshire Ramblers. In the main chamber, the party split. Calvert, Green and Cuttriss explored the passage to the south-east, now known as the Old East Passage, and the remaining two members explored the North West Passage, which led them to the West Chambers. Both passages were stalagmited, the first stalagmites to be seen in Gaping Gill, as the enormous main chamber is devoid of formations. They reached the surface again after six hours of exploration.

Calvert returned during Whitsun 1896. The whole of Saturday 23 May was taken in preparation and as usual with Calvert's parties, the shaft was entered by the Jib Tunnel Passage, now better known as Telephone Passage. This is the passage which houses the telephone cable connecting the surface and the floor of the main chamber on what are known as winch meets. On Sunday 24 May, Calvert, Booth and Elliet discovered Mud Hall, while Green and Cuttriss surveyed the main chamber. By the end of their explorations the team had added a number of passages and chambers. On 25 May, the alpine climber Slingsby, well known for his Norwegian mountain ascents, visited the cave and a further day was spent in retrieving equipment from the cave. Calvert and his party produced the first comparatively detailed survey of the system as known at that time.

Although Martel's and Calvert's explorations were followed by constant visits to the cave by various parties, including the first lady visitors in 1906, no new outstanding exploration was made until 1908 when a group attached to the Yorkshire Speleological Association, a society then two years old, made the descent and discovered a worm from the moorland surface in the South-East Passage and a draught, which indicated an alternative way into the cave from above. An entrance was found on the surface, but it was a year before the way could be cleared into the system. This was an easier route to the main chamber and an alternative access and exit way when the beck was in flood. The group from the Yorkshire Speleological Association was equipped with ladders of cotton rope and when they came to ascend, the ropes had become so wet that the cotton stretched as they stood on the rungs, so much so that they had to tread forty of the rungs before they could get off the floor, an amusing sight, but a tiring performance for the climbers.

In 1909, Blackburn-Holden and Simpson tried to find a third route into the cave by entering another hole on the moor, just south of Gaping Gill, but within a short distance they were barred by water and for this reason, the hole bears the name of Disappointment Pot. Contact with the main system was in fact made from this pot in 1944 by Leakey and his party, but only after overcoming a number of problems of clearing, water, tight places and several vertical pitches was the Gaping Gill system reached. This was certainly a very arduous alternative route. It joined the system in Hensler's Stream Passage discovered earlier in 1937 when the British Speleological Association, set up to co-ordinate cave records throughout the country, happened to have its annual conference in Yorkshire. Among the various alternative visits arranged by the Association was participation in a 'final' survey of Gaping Gill. Eric Hensler, one of the more distant visitors to the conference, chose Gaping Gill from the list of events. During the survey, he decided to crawl into a one-foot-high side passage. He continued for a quarter of a mile, much of it in a flat position, dragging himself along, until he came to a 60-foot-long chamber, rising to a height of 200 feet. From this chamber, he entered a system of stream passages, which intersected, forming a large chamber.

In 1948, the Brindle brothers, members of the Craven Pothole Club, opened Car Pot on the moor, on the opposite side of the footpath to Trow Gill, not far from Gaping Gill, but the passages were narrow. They got within six feet of the Gaping Gill system to find that the connection was only 2 inches wide, In 1948, too, the British Speleological Association found a connection from Bar Pot, which adjoins the wall skirting the path, leading to Trow Gill. The pothole contained two ladder pitches, the first of 45 feet, reached by a tight crawl downwards from the entrance, and the other 110 feet in the vicinity of some tight crawls, one of which leads into the South-East Passage of Gaping Gill.

No further discoveries of any magnitude were made until twenty years later, when in 1968 the Bradford Pothole Club, after much clearing, extended the system with the Whitsun Series. Practically every section or feature of a cave is named for easy identification. Sometimes they are named after the discoverer, as, for example, Hensler's Stream Passage, but this is difficult where more than one explorer is involved. Others are named after compass points, such as Eastern Passage, and others from their position in the system, Lower Series, Upper Series, and so on. Sometimes the nature of a passage or chamber may give rise to its name, such as 'Mud Chamber', 'Stalactite Chamber' or 'The Squeeze', but these may be repeated many times. Imaginative names are more distinctive, such as 'Primrose Path' for a particularly nasty crawl, or 'Tie Press' for a tight squeeze. The Bradford Pothole Club wisely tried to find an unusual name for a passage in the Whitsun Series. Why not coin a name, they thought, so they put all the initial letters of the members of the exploration team together and tried to make an anagram. No acceptable word was forthcoming, so they called the passage 'Anagram Passage'.

Important discoveries, especially after a long dormant period, give impetus in further explorations. Only a week after the Bradford Pothole Club discovered the Whitsun Series, the University of Leeds Speleological Association extended the Gaping Gill system even further by adding a series of passages and two chambers known as Clay Cavern and Mountain Hall, collectively known as Far Country. Even this was not the end of the discoveries, as in 1971 Wooding added another 1,200 feet, named Far Waters. The objective of most explorers of Gaping Gill, including Martel, was to link the system with the emergence of the stream at Ingleborough Cave, but a short link still remains to be made.

The show cave visitor, who regrets that he cannot experience the same excitement as the sporting caver, can actually descend the most famous pothole in Britain in safety, for twice a year there is a winch meet at Gaping Gill, one organized by the Bradford Pothole Club on the Spring Bank Holiday and the other by the Craven Pothole Club on August Bank Holiday, when visitors are let down the famous shaft to the floor of the Great Chamber in a winch-operated chair, for a modest fee, in the care of members of either of these two well-known British pothole clubs—and of course brought up again. If you do not fancy a winch meeting, it is still worth the walk up from Clapham to Ingleborough Cave and thence up the path through the impressive gully of Trow Gill to the open moorland and the entrance of Gaping Gill. Beyond Trow Gill, you pass Bar Pot over the wall on your left. A short distance beyond Bar Pot, you cross the wall, but be careful of open pots which are not protected. Gaping Gill is the one with the fence round the top. Passing through the gap in the fence you go down some earthen steps made in the side of the funnel of the pot and below you is Fell Beck, if in spate, tumbling down into the chasm which is Gaping Gill. The sight of the river tumbling over the edge is indeed spectacular and the noise as it falls the 360 feet can be terrific.

When talking about Gaping Gill, it would be difficult not to refer to the emergence cave of the system, Ingleborough Cave or Clapham Cave, as it is sometimes called. It is a show cave, but there is no public car access. The cave can be reached on foot from Clapdale Farm or, for a very small fee, you can walk by the pleasant lakeside path through the landscaped grounds of the Ingleborough Estate, a walk of about twenty or thirty minutes from Clapham village. Before

1837 the cave could only be entered for a short distance to a place where the passage was barred by a five-foot-high rock barrier, known as the Bay, damming a lake beyond with the water reaching almost to the roof. In September of that year, under the instructions of Mr Farrer, owner of the Ingleborough Estate, the barrier was removed and a considerable stretch of passage was drained. The remnants of the old rock barrier can still be seen in the sides of the show cave and beyond; at a height of four feet or so, the old water line can be seen. In 1838, a survey and plan were made of the cave and various details entered up by Mr Farrer in what was called the 'Cave Book'. The plan was revised during 1912 and 1913 by Charles Hill and Harold Brodrick to include a further portion beyond Giant's Hall, which was marked 'unexplored' on the early plan. At the time of the original survey Giant's Hall was known as Baron's Hall. The plan has been revised from time to time, but the original surveys were so well prepared that they formed a reliable basis for all future revisions.

Today, Ingleborough Cave is well lighted with a well-made path and guides. Beyond the point of the old barrier is an interesting cluster of stalactites known as the Inverted Forest. Instead of the usual pointed tips, they have bulbous ends, so that if turned upside down, they would look like trees. These stalactites have these peculiar ends because they were in contact with the surface of the original lake. The first chamber is Eldon Hall, named after the Earl of Eldon, who as Viscount Encombe was one of the 1837 explorers. Eldon Hall has a bank of flowstone which expanded laterally on the surface of the old pool to produce a mushroom-like formation. Ingleborough Cave is a good place to examine the effects on stalactites of contact with water surfaces. The 5-foot-long stalactite just beyond Eldon Hall, known as the Sword of Damocles is rather like the Dagger in Dan yr Ogof caves in Wales. Passing many formations, the visitor crosses a bridge over a couple of water pools to the Pillar Hall, where there is an enormously wide and 3-foot-high stalagmite, with a dimple in the top, known as the Jockey's Cap because of the flowstone peak formation on one side. From measurements taken in 1839 up to 1873, the growth of stalagmite seemed to proceed evenly forming $\frac{5}{17}$ inch annually, but since that time growth has been decreasing. Originally there was a stalactite above it, which in 1845 was 10 inches long, but it broke off when the measurements were being taken in 1853. In those eight years, it's length had increased by 4¼ inches, while at the Beehive, a stalagmite near the Sword of Damocles, there are 5-inch-long stalactites which developed in 150 years since the draining of the pool. All these examples show the great variation of stalagmitic growth and are a far cry from the old belief that it took a thousand years to form an inch. The drainage of the lake in Ingleborough Cave not only increased the length of the cave which could be visited, but produced good examples of the rate of stalagmitic growth and the effects on contact with the surface of the lake.

A stalagmitic column known as the Pillar gives its name to the chamber. Behind this, a natural pit takes away any accumulated water and the greater part of the flood water to join the water emerging from Clapham Beck Head by the path not far from the cave entrance.

A large passage connects Pillar Hall with Curtain Range where there is a series of curtain-like stalactites and examples of the results of stream action. Beyond the curtains, the roof lowers and you have to stoop for a few yards to Gothic Arch Passage, where you can stand up once more and proceed to the Long Gallery which is the end of the public tour, but not the end of the cave for cavers. Beyond, the Cellar Gallery leads through a muddy pool to Giant's Hall and the lakes Avernus and Pluto, which were found by diving along the connecting passage from Giant's Hall to Lake Avernus in 1953. Towards Gaping Gill, a system of passages leads by way of Far Eastern Bedding Cave, through the Wallows (water passages) to Inauguration Caverns and the Terminal Chamber which takes water from Gaping Gill. It may sound easy enough, but it involves mud and sandy crawls and a crawl through a stream passage, where the roof is just sufficiently high to enable you to keep your nose out of water, that is, if you do not churn the water about too much.

Just past the entrance of Ingleborough Cave, where the path bridge crosses Clapham Beck, is Clapham Beck Head Cave, where the water pours out to form the Clapham Beck, but the entrance closes down quickly inside. In 1951, an entry was made from another point, involving climbing down 12 feet of narrow fissure with a 12-foot crawl to the stream passage. Upstream there is a 14-foot waterfall and beyond this another 120 feet brings the explorer to a 12-foot-deep flooded pot from which the stream rises. The pot has been explored by cave divers, who have followed the totally submerged inlet passage leading to the bottom of it for 150 yards, but did not reach any dry passages.

Certain of the potholes round Gaping Gill have been mentioned in connection with attempts to find alternative entrances to the system, but there are others in the immediate vicinity, although leading to comparatively small systems. There is Nettle Pot, about forty yards south of Car Pot and Jean Pot south-east of it, but they merely lead into fissures. Marble Pot, about half a mile east-north-east of Gaping Gill, is a little more interesting and was first descended by the Yorkshire Ramblers in 1905. A stream flows into a miniature valley and down the pot at the end by a series of steps, followed by a 25-foot ladder pitch. Then comes a narrow crawl to a 60-foot pitch ending in a double chamber. There are other small chambers which involve climbing and a roof traverse. Sixty feet south of this pot is V. J. Hole with sheer peat sides, a particularly muddy place. It has two pitches ending in a chamber containing a thick mixture of water and peat. Rift Pot and Jockey Hole are due east of V. J. Hole in the same area known as the Allotment. Again, these pots have no known extensive systems. Rift Pot is the deeper, with five ladder pitches. Long Kin Cave and Long Kin Pots are other sporting caves in the Allotment area. Pots of this type are numerous in the Yorkshire moorland and it would need a great deal more than a chapter in this book to describe them, for the caving area of Yorkshire is very large and is one of immense caving activity.

A favourite pothole is Alum Pot, once known as Heln Pot, near Top Farm, Selside. Whether you call it a pothole or cave does not really matter, because you can either go potholing down Alum Pot or go caving by entering through Long Churn Cave, so it is both rolled into one. The pot was first descended by Birkbeck, who made the two unsuccessful attempts at Gaping Gill, but was successful in many other potholes. The entrance to the shaft is impressive, being 130 feet across and 40 feet wide, and today the exploration is often a combined operation of ladders and motor winch with bosun's chair. The winch is generally used for the 180-foot lift from the enormous rock bridge in the shaft of the pot. The bridge connects with Long Churn Cave. From the Bridge to the bottom of the pot is about 112 feet. The descent to the Bridge from Long Churn Cave is by 45 feet of ladder down the Dolly Tubs. The cave has other vividly descriptive titles to its various features, such as Double Shuffle Pool, Plank Pool, Dam Pool and Letter Box.

For those who like caves but have no wish to go potholing, there are other show caves as well as Ingleborough Cave. There are Stump Cross Caverns at Greenhow Hill, Pateley Bridge, on the B6265 from Grassington to Pateley Bridge, and White Scar Caves, a little over a mile from Ingleton on the Hawes Road. At Stump Cross the visitor is allowed to wander through the well-lighted cave passages without a guide so there is no waiting. You just buy your ticket and descend the flight of steps to the cave below. Like several of the Derbyshire caves, it was first found by miners seeking lead in 1858 or 1860. Near the bottom of the steps, a door blocks the way to a lower series, where Geoffrey Workman stayed in isolation from 16 June 1963 until 29 September 1963 and established his world record, his only link with the outside world being by telephone.

A few yards along the lighted passage is a side passage, almost cut off from the main passage by numerous formations known as the Butcher's Shop. Above it on a crevice are some fine flowstone formations giving the impression of a frozen waterfall. Further along the passage is a double stalagmitic column, the Twins, and further still we reach the Jewel Box, a kind of small fairy grotto with formations and a calcite-filled pool. After passing pure white deposit known as the Snowdrift, we come to Jacob's Well, a very attractive gour or rimstone pool and a realistically shaped Hawk stalagmite. complete with eye, hawk-shaped beak and tail feathers. Just beyond is the Sentinel, a fine stalagmitic column, and then comes the richly decorated

Chamber of Pillars and then the Sandstone Grotto with rimstone pools. The path goes up to the last and largest of the chambers, the Cathedral, which even includes a stalagmitic Organ. The roof of this chamber is adorned with straw stalactites and helectites. In the cave there is an exhibition of reindeer bones.

Like Stump Cross Caverns, White Scar Caves, which derives its name from the white limestone outcrops, adjoins the road. It was discovered in 1923 by a Cambridge undergraduate, Christopher Long. The original entrance was by a very narrow crack and so a tunnel 600 feet long was blasted through the rock to the well-decorated First Waterfall chamber. Much of the cave follows the present stream passage and paths have been laid alongside the stream and over it in places, so that the visitor can view the variety of formations without the risk of wet feet. These formations bear such names as the Madonna and Child, Elephant's Head and Buddha, but, in spite of these fanciful titles, the management of White Scar Caves has made a special point of bringing to the attention of the public the scientific aspects of the cave. Under the present management, as well as being a place of natural interest and beauty, the cave is a cave laboratory.

At White Scar Caves, the caver and cave diver are welcome participants in the search for speleological knowledge and have been successful in extending the known system beyond the show areas, even as late as December 1975, and it is possible that even more discoveries will have been made by the time this book is published.

The management is anxious that the public should be able to see such new discoveries and so, as well as improving the present facilities, they have embarked on a long-term plan to provide public access to these areas, including the Battlefield Chamber, which was discovered in 1971, an enormous cave chamber some 300 feet long and 50 feet wide and rising to a height varying from 20 to 40 feet. At the moment this chamber has only been seen by cave explorers, because of the difficulties of reaching it and the great jumble of boulders which it contains.

Apart from new exploration, various other branches of

speleology are being pursued at White Scar Caves. White Scar is a resurgence cave and a classical example of cave formation. The waters which flow through the cave originate as streams, flowing over the Yordale shales, forming the upper part of Ingleborough. As soon as the streams flow beyond the edge of the shales and meet the Great Scar limestone, they go underground through swallets. One of these is Boggart's Roaring Hole in Dowlass Moss, while further down the slope in the hollow of Crina Bottom, Greenwood Pot, a long, deep fissure, adds its quota as does the nearby Red Gait Sink. Even these do not account for all the water which enters White Scar Caves. Over thirty streams contribute to the volume of water in the caves in addition to surface drainage.

Where the stream leaves the cave at its exit, near the entrance to the show cave, it reaches the base of the overlying limestone and the last part of its journey is over the underlying slate of the Ingletonian screes.

It is to Yorkshire that we owe the introduction of the cave club, for the Yorkshire Ramblers' Club was the first caving club to be formed in Britain and that was in 1892. Although the club was intended for fell walking and climbing, it began to concentrate almost immediately on potholing and, like the Mendip Nature Research Committee, which followed it in 1906, then as the Mendip Nature Research Club, it is still in existence. The Yorkshire Ramblers' Club produced the first cave club journal in 1899. The Yorkshire Speleological Association was next formed in Yorkshire in 1906, but in spite of some very good work, the society was short lived. The Gritstone Club, which is still active, was formed in 1922, and in 1929 the still existent Craven Pothole Club was started. The Northern Cavern and Fell Club came into being in 1928. About 1935, the year of the inaugural meeting of the British Speleological Association, came the Bradford Pothole Club, still very active, and in 1936 the Leeds Pennine Club. Since that time, many new clubs have been formed, as is the case in other regions. Some of the clubs mentioned had even earlier beginnings than their inaugural meetings, having developed from earlier and sometimes nameless groups of cavers.

EIGHT

Caves of Derbyshire

IF WE DRAW a line on a map just outside Castleton, Bradwell, Calver, Bakewell, Matlock, Wirksworth, Waterhouses, Buxton and back to Castleton, we would enclose the area of Carboniferous Limestone and consequently the caving area of Derbyshire. However, within this area is a small part of Staffordshire, which includes the entire Manifold Valley and its caves on the boundary of the two counties. Although the far larger mass of caves is in Derbyshire, it would be wrong to omit those of the Manifold Valley and so the whole area is often called the Peak District caving area.

Derbyshire is a mixture of caves and mines and sometimes both in one, for Derbyshire was probably the most important lead-mining area in Britain in times when it was a flourishing British industry. A number of mines broke into cave systems, while some caves produced lead for the taking. Not only was lead mined, but other minerals too, and of course the famous Derbyshire Blue John stone, 'the Peakland Jewel', with its banded colours and translucency, has for many generations been an attractively veined and decorated material for making bowls and vases.

The first organized cave club in Derbyshire was the Derbyshire Pennine Club, which was inaugurated in 1906, although it had grown out of the Kyndwr Club. The beginning of the century and particularly round 1906 seems to have been the period of establishment of our older cave clubs. These were the years immediately following Martel's success at Gaping Gill and the oldest of all the clubs, the Yorkshire Ramblers, preceded that event by three years. The Peak had a good start in the discovery of caves, for the 'old men', the miners, had found a lot of them long before the establishment of speleology in the district, as a sport and a science. It is for this reason too that the Peak has such a high

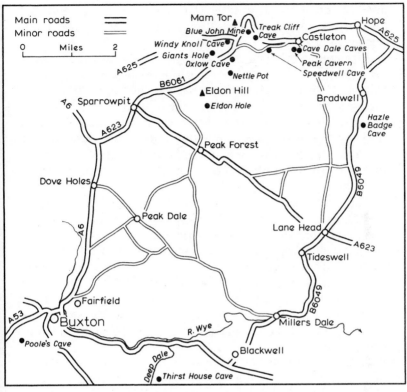

BUXTON/CASTLETON CAVE AREA, DERBYSHIRE

quota of show caves, mostly conveniently situated in two groups at Matlock and Castleton. Miners found caves in other regions, in Yorkshire and Somerset, but most of these caves were not so conveniently placed or so suitable for the public to visit. In Derbyshire too, the mining activities were more extensive and of longer duration for the lead was regarded as of superior quality when compared with the Mendip lead and justified the introduction of new techniques and greater output. The results have been advantageous to cavers and commercial cave operators alike. However, none of this alters the fact that the Derbyshire caving clubs have done a great deal during the caving era to add new discoveries and knowledge to the not so gentle sport and science of caving.

Although Yorkshire and Derbyshire are adjoining counties, there are marked differences between the caves. Yorkshire is famous for its many potholes, gaping shafts on the high moorlands with becks tumbling a hundred feet or so into chasms. There are shafts in Derbyshire, but many are those sunk by the lead miners, and there are a few potholes, but generally the cave systems do not run to a great depth with the numerous internal ladder pitches common in many Yorkshire caves.

There are some notable exceptions to this rule. Nettle Pot at Castleton was first explored in 1935 by the Derbyshire Pennine Club, who had to clear the top 90 feet of debris. The pot has four ladder pitches, varying from 35 to 180 feet. The top part of the shaft is very narrow and care has to be taken to avoid dislodging stones during the descent. The total depth of the system is 520 feet. Eldon Hole, near Nettle Pot at Castleton, was first explored in 1772. It is not so deep as Nettle Pot, 180 feet to the lowest part. The shaft ends in a side chamber.

Oxlow Cavern at Castleton is on the side of a hill about a quarter of a mile from Oxlow House. Like many caves, visiting is controlled and the cave is locked. There are five pitches. The first pitch is at the entrance and involves about 55 feet of ladder. This leads to a chamber, some 30 feet long and 30 feet wide, with a passage connecting with two pitches, which take the caver to a second chamber, known as East Chamber. A fourth pitch is reached through a small hole and then there is some scrambling to arrive at the final chamber and the last pitch.

Apart from the crowds on 29 May to see the Garland procession, many visitors go to see one of the most impressive sights of Derbyshire, Peak Cavern. The great entrance of this cave with the ruins of Peveril or Peak Castle of Sir Walter Scott's novel *Peveril of the Peak* above it dominate the village.

The cave, now a show cave, is seen as an enormous opening in the hill and is approached from the village by the Peak

Gorge, which ends at the cave entrance. So large is the mouth of the cave that it once housed a whole community of rope-makers, as the huge sheltered space of the vestibule gave them ample room for their rope walks. The cottages they built under the shelter have long since disappeared and the last of the community, Mary Knight, died in 1845, although occasional rope-making has been carried out there since that time. Relics of that community still survive in the rope-making equipment to be seen in the cave entrance, the largest in the British Isles.

Inside, the cave is not so spacious. Indeed you have to stoop for a few yards just inside the entrance and this part is known as Lumbago Walk. A short way along is the Inner Styx, a long but shallow pool which once used to be crossed by lying flat in a punt to avoid hitting the ceiling, one guide walking in the water pushing the punt and another beyond the arch pulling it. A dry way has been made by blasting a passage along the rock wall, a facility provided for today's visitors who proceed more comfortably than did Queen Victoria on the two occasions when she visited the cavern and had to go by punt. Byron also made the journey in this way accompanied by Mary A. Chaworth, who inspired the poem 'The Dream'. Frankly, I think it would be more exciting to go by punt and you can still experience this form of cave travel in Speedwell Cave. Beyond Lumbago Walk we come into the 98 feet to 100 feet high Great Cave. A shaft in the ceiling, explored in 1902, connects this chamber with the surface near Peveril Castle, but its top has been concreted over. At the end of the Great Cave is Roger Rain's House, not named after a person but because the 80-foot-high chamber rains water from the roof at one spot. High in the roof is a passage which connects with the Great Cave. This opening is called the Orchestra Chamber or Chancel, where choirs have sung on occasions, as at Wookey Hole, Somerset. Such a performance was provided at Peak Cavern for Queen Victoria's entertainment. Apparently such performances were quite frequent, for the Reverend R. Ward in his *Guide to the Peak of Derbyshire* wrote in 1827 "that the stranger is generally surprised by a concert" and experienced mixed feelings of "fear and pleasure, astonishment and delight". After the performance, the eight or ten women and children would appear in the rock hollow to which they had previously clambered about fifty feet above the visitors.

From the Great Cave, the path goes through Pluto's Dining Room to steps leading down to the path which follows the stream through the Five Arches, an interesting feature of river erosion to Victoria Cavern where the roof is known as Great Tom of Lincoln because of its bell shape. The show cave ends here, although the cave goes on by narrow passages and deep pools to connect with Speedwell Mine. Water from the mine flows through Peak Cavern to Russet's or Russett's well by the bridge outside the cave. This was once the water supply for the village.

Speedwell Mine or Cavern, formerly known as Speedwell Level or Navigation Mine, is but half a mile from Castleton on the road to Winnat's Pass, on the same side as Peak Cavern. An upward path from the road leads to the entrance in what is known as Long Cliff. The mine owes its origin to a group of Staffordshire miners seeking lead and who worked there for some years with little success. It is not known exactly how long mining was in progress at Speedwell as accounts vary from seven to eleven years.

Today, as the visitor is propelled by boat through the narrow and low tunnel of the mine, it is not hard to imagine the old miners who had to make it and this took something like four years. The guide points out lead veins in the wall of the tunnel. It is hard to believe that the amount of lead obtainable would have warranted the expenditure of time and energy making the man-hewn tunnel through which we pass, but the miners were looking for much larger deposits and larger veins than those like the Poor or Little Winster Vein which the boat passes. Further along the tunnel is the Long Cliff Vein, obviously a more productive vein, as the miners are reputed to have followed it for half to three-quarters of a mile. Much of it is not accessible today and is sealed off because of the unsafe nature of the workings. The driving of the Speedwell Tunnel was not just 'a shot in the dark'. The work was based on a knowledge of existing rakes from other mines and surface prospecting. The object was to intersect the rakes, but the chance taken was whether they would be economically productive. At Long Cliff the rakes run east to west and hence the workings in the cliff were driven southwards and, as we see, the tunnel crosses them more or less at right angles.

Today, we descend into the mine by 104 steps and below a boat taking twenty people is waiting. There is a guide at the front and another to propel the boat at the back by leaning against a supporting frame and walking the ceiling, as was done by canal-boat operators when passing under low tunnels in the days of horse-drawn boats. The guide in front helps to guide the boat with his hands by keeping it away from the sides and gives his passengers a description of the mine. At one point, he stops talking and brings the boat to a standstill, but instead of the silence you expect underground, there is a peculiar mumbling noise, rather like a number of low moaning human voices merging into an uncanny rumble. It is the distorted echoing voice of the guide of a returning boat. A little further along, the boat is turned into a bay at the side of the tunnel and waits for the returning boat to pass.

At intervals along the tunnel wall can be seen the straight grooves left by drilling, each about 18 inches along and over an inch in diameter. It was a slow and laborious process carried out by hand. One miner would turn the drill, while the other struck it and, as two holes were usually made at the same time, four miners would be engaged on the task. When fifteen or twenty holes had been dug, they would be 'blown', that is, filled with gunpowder and a fuse, sealed and ignited, very different from the automatic drills used in stone quarries today and the electrical firing from shelter at a safe distance. As there was no shelter, alcoves, like the ones often seen in railway tunnels to protect plate layers from passing trains, were hewn out of the walls and the miners had to squeeze into them as stones and blast would go along the tunnel as if from a gun barrel. The smoke that followed must have been thick and stifling and must have hung about the tunnel for some time, as there was no through ventilation. As the tunnel grew longer, the conditions grew worse and the miners were allowed certain days off. At one point, an alcove was made to accommodate a boy with a hand-operated bellows to help clear the fumes.

At the end of the long tunnel journey, the boat comes to a landing platform over a hole about fifty feet above the floor of a natural cavern. The cavern runs across the axis of the tunnel and is bridged over. The visitor can look down from the bridge to the sloping rocks below and the lake, with its "stygian waters" as the Rev. Wood described them in 1827. This is known as the Bottomless Pit. 40,000 tons of rubble from the further workings are said to have been thrown into the lake without making any difference to the water level. Cave divers have plumbed the depth of the water and found it to be about thirty feet from its surface to the top of the submerged stone fill.

The water in the tunnel is kept back by a sill in order to keep the tunnel from draining and to provide sufficient draught for the boats. Any surplus water pours over the sill into the pit below and produces a spectacular waterfall. The roar of the fall can be heard from the boat some time before reaching the landing stage. In the wall of the cavern can be seen a thick rake known as the Faucet or Foreside and the remains of the timber projections show that the rake had been worked. Like many rakes, it has comparatively little galena, most of the cavity being spar.

The show cave or mine visit is ended, but as the boat cannot be reversed, the resting frame is moved to the opposite end of the boat and the passengers about face, for what was the front for the outward journey becomes the back for the return journey.

Beyond the Bottomless Pit, the mine tunnel continues and is called the Far Canal and again joins natural stream passages. Exploring these is difficult and such names as Whirlpool, Boulder Piles, Assault Course are enough to deter any visitor out for a pleasant roam through a show cave and he would suspect that even the Bathing Pool was a little different from a pleasant dip—and he would be right.

Higher up the road to the Winnat's Pass and still on the same side as Speedwell Mine, steps lead up to Treak Cliff Cavern, which takes its name from the rock height adjoining it. The old name of the cave was Miller's Vein Mine, worked not for lead but for the fluorspar deposits. This was the first vein to be worked for that mineral. Originally the mineral was worked for ornamental purposes, but during the 1914–18 war it was worked for flux for blast furnaces used for the manufacture of steel. The winning of fluorspar continued at Treak Cliff for some years after the war and in 1926, the operators broke into a series of stalagmitic chambers, called the New Series. Both the Old and New Series were later converted to a show cave and opened to the public in 1935. A small controlled amount of Blue John, a rare variety of fluorspar is still taken from the workings, but is used only in the manufacture of vases, ornaments, semi-precious jewellery and other decorative objects.

Having passed through the entrance tunnel, the visitor is in the Old Series where the Blue John vein can be seen. The lights are passed through thin slices of Blue John to show the various translucent colours of the fluorspar and by a manipulation of lighting, the guide throws a silhouette of a witch riding her broomstick on the cave wall. A corridor links the Old Series with the New and there we find Aladdin's Cave with its many pretty formations. From there it is an easy step to Fairyland, a chamber with a number of carrot-shaped stalactites, then to Dream Cave with a fine cluster of slender pointed stalactites and a large collection of formations where the guide points out a stalactite and stalagmite which have only 1½ inches to go before becoming a stalagmitic column. There are various familiar shapes like the Elephant. The last time I was there, I wondered how many elephants I had seen in caves, whole or in part, hawks, parrots, pincushions galore and frozen waterfalls by the score. I thought that what was true of Charles Cotton's description in his seventeenth-century poem of one cave was true of another:

> Propt round with peasants on you trembling go,
> Whilst, every step you take your guides do show
> In the uneven rock the uncouth shapes
> Of Men, of Lions, Horses, Dogs and Apes:
> But so resembling each the fancy'd shape,
> The Man might be the Horse, the Dog the Ape.

The last chamber in Treak Cliff Cavern is the Dome of St Paul's, needless to say because of the shape of the ceiling.

The last of the string of show caves is Blue John Caverns, under the shadow of Mam Tor. The entrance to Blue John Caverns, like Treak Cliff Cavern, is by an artificial tunnel from which the old shaft can be reached. One of the first features to be seen is the Roman Level for traditionally, the earliest working of Blue John was by the Romans. As far as I know, no positively identifiable Roman artifacts have been found in the fluorspar mines, but the fact that two large Blue John vases were found in the ruins of Pompeii would indicate that the material was exported to the Roman world. There is, also, no evidence that the Romans did deep mining in this country and they normally obtained their minerals by surface excavation. Nevertheless, whatever method was adopted, it appears that Blue John was being obtained from the area in Roman times.

Blue John fluorspar owes its attraction to the various colours, and the high degree of mineral staining is very apparent in the Grand Crystallized Cavern which is entered by the gently sloping passage known as Ladies' Walk. The curiously ribbed and coloured walls and the crystals in the cave, now illuminated by electric light, were once shown by hoisting a twenty-eight candle chandelier into the dome of the ceiling by means of a rope, pulley and small windlass. The chandelier was described in 'The Gem of the Peak' by Hedinger in 1846, but it was lost until 1959, when it was found high up in a recess of the chamber. The windlass and chandelier are now exhibited in the caverns, with a number of other objects, including a long-handled flash pan in which a Bengal light used to be ignited for illumination. Some of the tools used for the mining of fluorspar can be seen, a small wooden iron-wheeled trolley used for carrying the stone, rather the worse for wear, and a cylindrical bellows for air-changing with its projecting ribs, looking rather like a corrugated metal oil drum. Such bellows were usually operated by a ten-year-old boy. Quite a few hand tools have been recovered and exhibited in the caverns.

The next chamber is the Waterfall Cavern, where the coating of stalagmite is found on one side of the chamber only. Then comes the Stalactite Cavern and the name is a good description. The next chamber of the tour is Lord

Mulgrave's Dining Room, a large circular void, with a name which comes from the belief that Lord Mulgrave held a dinner there for the miners. There was certainly no need for wall decoration, as large areas are covered with scintillating deposits and in a corner can be seen a Blue John vein.

The last chamber of the show cave is the Variegated Cavern and, as its name implies, there is a great variety of colours in its walls. For those interested in geology, Blue John's Caverns are a happy hunting ground. Small specimens of fluorspar can be bought at the caves, either unworked or set in jewellery. It is now recovered only in small quantities and the days when Blue John was used for the making of large vases such as those at Chatsworth House and the Vatican have gone. It is a fragile stone, easily broken and fluorspar, of which the rare Blue John is an example, can be found in a whole range of colours.

On Mam Tor, near Winnat's Head Farm, is Windy Knoll Cave, found during quarrying. This cave produced nearly 7,000 ancient animal bones, including bison. bear and wolf. The bones are now distributed among a number of museums, including the British Museum (Natural History), Buxton, Middlesbrough, Bolton and Derby.

In the 1870s a small cave in Cave Dale, below Peveril Castle, revealed a number of animal bones which had been broken as food bones, together with some crude pottery, chips of flint. a bronze axe and a perforated stone axe hammer. Although the latter would indicate Bronze Age occupation, it may be possible that some of the finds belonged to the New Stone Age.

South of the village of Bradwell, not far from Castleton, is Hazelbadge Cave, near Hazelbadge Farm and within two hundred feet of the road. The system is very small and is entered by a 25-foot-deep shaft.

Bradwell is better known for Bagshawe Cavern, originally found by four miners in 1806 during their search for lead. It can now be seen by the public, although arrangements have to be made in Bradwell for viewing. It is not developed as a show cave as are the caves at Castleton. There is no electric lighting or made-up paths and it is preferable to wear old clothes. In mining days, it was known as Mulespinner Mine.

A small stone building now conceals the entrance, but from the building steps lead down through a rift, which was cleared by the miners, to a passage made by the owner before the 1914–18 war. The passage connects a number of small caverns, which the guide illuminates for your benefit.

At the end of the passage, steps lead up to a small chamber and a left-hand passage takes you past some fine formations to the Dungeon. Although this is the furthest point for the public, more of the cave is accessible for the caver by descending a shaft in the floor of the Dungeon to a low chamber with a floor sloping to water. This is known as Blackpool Sands from which the Hippodrome Chamber and the Glory Hole Can be reached. It is said that Bagshawe Cave was named after a woman who explored it in 1830.

Bagshawe Cave consists mainly of three passages, more or less parallel but at different levels, and although the chambers are not comparable in size with those of the Castleton caves, it is an interesting system in that it conveys to the visitor some idea of a natural cave without the tourist trimmings. It is almost a textbook example of the process of cave formation, the upper galleries being the oldest and the water gallery the newest.

Another cave area is Matlock Bath. Like Castleton this is a most attractive locality, but while Castleton is surrounded by open hills, Matlock Bath is tucked away between the tree-clad slopes of the River Derwent. Matlock Bath is a delightful spa village strung along the banks of the river. Overlooking the village is Masson Hill with the steep wooded slope of the Heights of Abraham, laid out with easy ascending paths to its crowning feature, the Victoria Prospect Tower. It is worth making the easy climb to the top of the Heights for the view and the variety of its trees and wild flowers. You can sit on the veranda or in the beer garden and enjoy the view of the white buildings of Matlock Bath between the trees below and the dark silhouette of the castle on the opposite skyline. The sale of beer here goes back much further than the present Rutland Tavern. Beer brewed on the premises used to be sold to the miners from the window of the adjoining house and it is thought to be the oldest beer licence in Matlock.

It is on this hill that Great Rutland Cavern and Great Masson Cavern are to be found. Great Rutland Cavern, immediately behind the Rutland Tavern, was known as the Royal Rutland Cavern in the nineteenth century and it has been visited by royalty on several occasions. Not only did Queen Victoria visit Peak Cavern in 1834 and 1842, but in 1832 she visited the Cumberland Cave, which in consequence became the Royal Cumberland Cave, now closed, and Rutland Cavern. On her visit to the Heights of Abraham, she was entertained by the Bakewell Brass Band. Her liking for caves must have made them fashionable, for eight years later the Dowager Queen Adelaide visited the Heights, but having climbed to the entrance of the Rutland Cavern the *cortège* turned and went down the hill again, not, it is thought, from any sudden attack of claustrophobia, but pressure of time and a particularly tight programme. The Grand Duke Michael of Russia visited the Rutland Cavern in 1818. He apparently had more time to spare, for, on the way up, he sat backwards on his pony to face the ladies following him, so, as he said, he could pay them the highest possible compliment. None of these personages visited the Masson Cavern, as it was not opened to the public until 1844.

The earliest written reference to the working of lead in the area is in Domesday Book where we are told that lead was being worked in the Nestor Mine, the former name of Rutland Cavern in 1080, but how long it was being worked there before that date there is no evidence. We know that the Romans exploited Britain for lead, for they used it for various purposes, for example, the Roman baths at Bath were lined with it. The Romans also worked lead to obtain silver for coinage and articles, although many tons of lead have to be smelted to recover even a small quantity of silver. There does not appear to be any direct evidence that the Nestor Mine was worked in Roman times, unless the small hack marks of the older parts of the workings, smaller than those normally left by the medieval and later miners, are of Roman origin. The Roman lead workings were usually surface ones and one of the most important veins in Derbyshire, the Great Rake, crosses the Heights of Abraham and can be mined on the surface. As Rutland is a natural cave, lead could be obtained there too without deep mining, Also, the name Nestus may well be of Roman origin.

Great Rutland or Rutland Cavern, named after the Duke of Rutland whose seat is Haddon Hall not far away, is certainly a most interesting cave and enlivened by models of Roman soldiers and native miners carrying picks or hacks of the type used for the recovery of ore. A number of old mining tools have been found within the cave and the black marks left by the ignited bundles of reeds used as torches still remain as does the mouldering remains of a stemple, high in the roof, used for climbing. In the rock walls, a variety of minerals can be seen within a few square yards. Specks of bright green malachite or copper ore occupy small holes in the rock. There are toadstone nodules, bands of ironstone, the black galena of the lead miner, and the colourful fluorspar, while the surface glistens with the gold flecks of iron pyrites, 'Fool's Gold'.

The Great Rutland and Masson Caverns may lack the quantity of stalagmitic formations to be found in some other caves, but make up for them by their mineral wealth and a feeling of contact with the old miners.

The Great Masson Cavern or mine does not have the electric lighting of the Great Rutland Cavern and not even the old Victorian gas lighting system which once existed in Great Rutland. The visitor depends on the hand lamps for viewing this cave which is reached by walking up the path from Rutland Tavern, towards the Victoria Prospect Tower. The system is entered by an open top incline becoming an underground corridor, leading to the natural cavern, as like many of the Derbyshire caves it is a combination of natural cave and mine workings. Although the public portion is about 400 yards, there is a further half-mile system of old workings leading to King's Mine and Knowllys Mine.

Opposite the Heights of Abraham is High Tor where there are a number of open clefts and fissures in the surface, known as Fern Cave and Roman Cave. These were surface workings for lead in the Great Rake, which crosses the Heights of Abraham and is continued on the other side of the Derwent.

Just over a mile north-west of Great Masson Cavern, near Leawood Farm, is the system known as the Jug Holes or Jug

Hole Cave in which can be seen volcanic intrusions known as 'toadstones' as in Great Rutland Cavern. During a survey in 1948–51, it was found that Jug Hole Cave was the only cave habitat of the Whiskered Bat known in Britain. The system consists of shafts, natural voids, a bedding plane angled at 45 degrees and a number of mining levels.

Derbyshire is not only remarkable for the number of its existing show caves, but also for those which are now closed to the public. At Matlock there have been several show caves which are no longer open, while at Buxton there is Poole's Cavern, again an attraction of the town. It is said that it took its name from an outlaw who lived there in Henry IV's reign. One of the chambers was known as the Roman Chamber and the cave produced Romano-British and other relics. In this chamber, the guide would point out stalagmitic formations in the shape of familiar objects, many of them associated with the outlaw, such as the saddle of his horse. He seems to have possessed such unlikely articles as a woolsack and a turtle and there is the usual stalagmite curtain, the flitch of bacon which he never cooked. Then there is the stalagmitic pillar against which Mary, Queen of Scots, was supposed to have leaned when on a visit to the cave. It is said to mark the extent of her tour and beyond the progress became difficult.

Although the Manifold Valley is in Staffordshire, it is so close to the Derbyshire border that some reference should be made to it. It is a favourite walk for local people and visitors alike. All along the valley, rocky bluffs add scenic interest and contain cave entrances. The most spectacular of these is the impressive 30-feet-high entrance of Thor's Cave in Wetton Low, overlooking the river about 250 feet below. Like a number of the caves in the Manifold, it has produced archaeological material. The cave entrance is not difficult to reach and in fact the smooth rock floor inside can be more difficult than the approach. It is a well-lighted single chamber, as a second fissure-like opening on the side illuminates the back of the cave. On one occasion when I visited the cave, a potholer was suspended in a harness from a piton driven into the ceiling and in a horizontal position was driving in another piton above his head. He transferred his position to the new hold, while he drove another piton into the rock crack. A line of pitons showed that he had travelled some way along the ceiling already and there he was, suspended like a fly over twenty feet above the rock floor. When he eventually came down, we sat on a rock and chatted about caves. The subject came round to my own diving activities at Wookey Hole. "I wouldn't like to do that," he said, "far too dangerous." I thought I would feel much safer in a diving suit on the bed of the Axe than up there suspended from the ceiling. I suppose it is a matter of opinion.

Thor's Fissure Cave is close to Thor's Cave. Although the entrance is wide, it narrows down to a fissure inside. At floor level, it is only between 2½ and 5 feet wide and about 60 feet long. It may be a continuation of Thor's Cave, but there is no penetrable connection. The cave can be approached through the west window of Thor's Cave, but it involves a difficult traverse and climbing, while the direct route from the base of the cliff can only be taken by an experienced rock climber. An easier alternative is by a steep grass descent from the south end of the top of the cliff, but it should be done only with somebody who knows the way. In spite of the difficult access and narrow working space, the cave has been excavated by a team of cave archaeologists. It is a good example of the problems often experienced in excavating a cave site and I have always believed that excavators of cave sites should know the nature of caves and cave exploring. The finds from Thor's Fissure Cave were numerous and important, representing several periods. The animal bones included deer. reindeer and bear and there were remains of several human skeletons. There was an amber button and beads of the Bronze Age and pottery and articles of Romano-British times.

None of the caves in the Manifold Valley can be said to be extensive systems. They are mostly shallow caves, but a reasonably complicated cave is St Bertram's Cave in Beeston Tor. The lower and upper entrances to the cave can be seen from the path adjoining the Manifold on the opposite bank of the river and if you wait there long enough you will usually see a caver disappear into the bottom entrance and, a short time later, appear at the window higher up the cliff.

North from Thor's Cave can be seen Hannah Woman's

Cave, or Old Hannah's Hole, a 10-foot-high fissure, but like many cave entrances it quickly narrows down. Some human remains were excavated there in 1896. Above it is a gully where some peculiar explosions occurred about that time and they have been reported on a minor scale since. In 1896 they were witnessed by a number of people on the same occasion and no one could give an explanation, except that they occurred when there were very strong winds. A Mr Fallows and a Mr Wint were driving a couple of cows below the gully when a blue flame tipped with reddish yellow suddenly emerged from a cleft, accompanied by a terrific explosion, causing the cows to run off with their tails in the air. When the men reached Wetton, some of the villagers went off to a viewpoint from where they could see flashes of fire, accompanied by blasts. Mr Wint said he compared the noise to a falling building and that the blue blaze appeared to be about twelve inches broad. At the same time there was "a noise like the crackling of a forest fire".

When Sir T. Wardle and G. Barrow of H.M. Geological Survey were near the fissure in 1898, they suddenly heard a noise like several rifles fired in quick succession, followed by an explosion and flash which emerged horizontally from a crack in the rock. A short time later a larger explosion took place accompanied by a six- to eight-inch band of "bluish vapour".

Another area of small caves containing archaeological material in Derbyshire is Creswell Crags, on the border with Nottinghamshire. Creswell Crags gives its name to the latest of the Old Stone Age phases of Britain, Creswellian, because of implements of this period found there. The important caves are Mother Grundy's Parlour, Pin-Hole, Robin Hood's Cave and Church Hole, the last named being in Nottinghamshire on the other side of the valley. All are fissure caves and were systematically excavated between 1875 and 1879. In Mother Grundy's Parlour and Robin Hood's Cave were found bones of hippopotamus and rhinoceros which had been gnawed by hyaenas. In fact all four caves had been hyaena dens and all contained deposits of the Old Stone Age. From Robin Hood's Cave came an engraving of a horse's head, the first piece of Old Stone Age art to be found in Britain.

Although the initial excavations at Creswell Crags were carried out in the last quarter of the nineteenth century, further important excavations were begun in 1923 by Mr A. Leslie Armstrong who found in Pin Hole Cave, among other important material, a piece of bone with a human figure engraved on it. There are other caves and rock shelters at Creswell Crags and in one of them, Whaley Rock Shelter No. 2, Armstrong found the skull, with four holes, of an Old Stone Age woman, who appeared to have died a violent death.

NINE

Caves of South Wales

THE PRINCIPAL CAVING AREA of Wales is in the Carboniferous Limestone uplands of South Wales lying part way between the heights of the Brecon Beacons massif and the South Wales Coalfield. The Carboniferous Limestone reappears in the Gower Peninsula west of Swansea and in West Wales. There are a number of minor caves in North Wales, but it is not proposed to deal with these in a general book of this nature. Those interested can refer to the bibliography.

The first caves of South Wales to attract attention were those in the head valleys of the River Neath. On the Mellte, Porth yr Ogof has, over the centuries, attracted the attention of those in search of natural curiosities. From time immemorial, its large archway fascinated travellers, poets and artists. The more daring of writers even penetrated into its 'stygian' depths, often with the aid of local guides. and the stories they told of the distances they covered and the wonders they saw were much exaggerated. This was not only true of Porth yr Ogof but of many early descriptions of caves. The modern caver will, however, often forgive their stories, as in the light of flares or candles, shadows thrown can play strange tricks, and depths and heights seem greater than they are. 'Bottomless gulfs' become insignificant under electric lights, and vast chambers shrink. Distances underground always seem much longer than they are because of the difficulty of progress and the limitations of assessing distances in darkness.

When Michael Faraday visited it in 1819, the area round Porth yr Ogof must have seemed remote, as indeed it was when I first visited it. Today, it is not unusual to find traffic congestion in the lanes round the cave as minibuses bearing the names of universities and schools wind their way to the

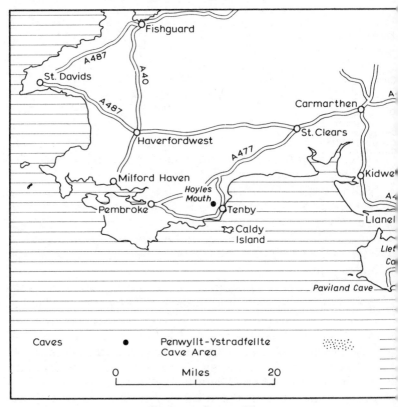

CAVES OF SOUTH WALES

huge car park provided by the Brecon Beacons National Park. The cave entrance is 50 feet wide and 15 feet high; lecturers from near and far stand on suitable rocks and pour out information about the various geological features to groups of students armed with notebooks, while in the cave itself, voices can be heard from the parties of schoolchildren and others negotiating passages in the cave under the leadership of school instructors. So popular is the sight of Porth yr Ogof that wooden seats have been provided on the steep slope of the glen for the benefit of the old and infirm and warning notices have been erected, drawing attention to fatalities which have occurred within the cave.

Like other Welsh cave systems, Porth yr Ogof is partly a

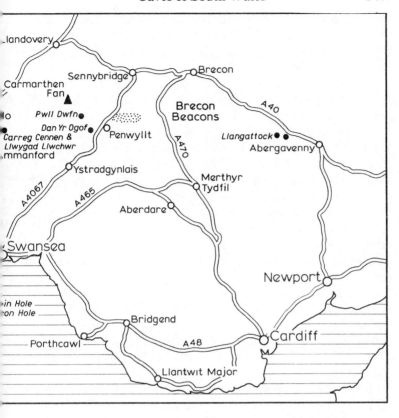

wet cave and the River Mellte, like other cave rivers of
Wales, can be very temperamental, sometimes slow running
or even absent in places and at other times running full spate,
since it is fed from sources high in the mountains. It can rise
in a surprisingly short length of time. Even if you do not go to
Porth yr Ogof for caving, it is worth visiting for its setting is
in an area rich in waterfalls and wooded ravines. The Mellte
River is but one of the rivers which have cut their ravines in
the immediate neighbourhood, producing caves and, down
river, waterfalls. Like all of them, the Mellte has its charms.
In dry weather, the course of the Mellte, immediately outside
the entrance of the cave, is marked by a dry boulder-strewn
river bed, for in this area it is an intermittent river, taking an
underground course, and as with sections of neighbouring

underground rivers, you can hear the stream running below. When the underground course becomes overladen after rain, the river rises above the boulders and, in full spate, enters the cave above the surface.

The spot at which the river goes underground north of Porth yr Ogof varies according to the height of water. The drier the conditions, the nearer is the point of engulfment to the village of Ystradfellte (Vale of the Mellte), where Michael Faraday once stayed on holiday. Although the river may be absent at the entrance, it reappears a little distance within the cave as a large pool. If the visitor ventures into the cave mouth and picks his way across the boulders, the edge of the pool can be reached. There is a glimmer of daylight for most of the way, but it is preferable to take a torch as the going is rough and slippery. On the far side of the pool, a composite calcite vein shows up against the surrounding darkness and, with a bit of imagination, this looks like a horse. It is this feature which has given the cave its alternative name of the Cave of the White Horse. At the far end of the cave, the river flows out through a narrow slot into the open air. Just within the entrance and on the right is a maze of dry passages, but these can only be explored with the aid of a light.

The main course of the cave can be traced from the surface without entering the cave. If you cross the road by the car park, a footpath follows a series of depressions and one or two are actually open, giving access to the river below. Finally, the footpath overlooks the exit or resurgence of the river, which is an impressive sight, as the Mellte flows through a narrow cleft between water-smoothed rocks, to open out once again into a broad leafy valley.

Not far to the west of the Mellte lies the Little Neath or the Nedd Fechan, with its small, narrow cave system in the area of the ford Pwll y Rhyd. In the mid-1930s parties of cavers from Bristol started to visit this area to explore such systems as the White Lady Cave, Bridge Cave, Town Drain and Ogof Pwll y Rhyd, combined with a visit to the nearby Porth yr Ogof.

In 1937 attention was turned to the valley of the River Tawe to the west of the Neath, for that year saw the first expeditions of Gerald Platten and his group to the cave system of Dan yr Ogof, now partly a show cave. The cave was not so accessible as today, for its only entrance was by way of the River Llynfel, which emerges from an arch in the rock and, after tumbling over a fall, one of today's tourist attractions, finds its way into the Tawe. The Tawe eventually flows into the Bristol Channel at Swansea, or to use the Welsh name, Abertawe (Mouth of the River Tawe).

It was this difficult access which explains why the cave had not been entered since the original exploration by the Morgan Brothers. When Jeffrey and Ashwell Morgan first entered the cave in 1912, they found their way into a large system of dry passages and chambers until they came to a series of underground lakes which prevented them going any further. The next day, they returned with a third brother, Edwin Morgan, and their gamekeeper, Morgan Richard Williams. This time they crossed the first lake by means of a raft. Their sole illumination was that of candles and, in the sand of the cave floor, they fingered arrows to help them find their way back.

On the third day, a fourth member of their party, Mr William Lewis, joined them, when they marked their way by night lights. It was on this occasion that they placed a thermos flask and a piece of paper containing the names of the party in a bottle, which was found in place during the 1937 investigation. Across the first lake, their way on was barred by a second lake and so they acquired a coracle from Carmarthen and made a further assault with this frail craft. Mr T. Ashwell Morgan managed, with outstanding tenacity and courage under these primitive conditions, to cross two further lakes and reach the foot of the waterfall, which was the limit of their discoveries.

By 1937, the brothers had purchased the site and invited a Yorkshire Ramblers Club team, led by Mr E.E. Roberts, to explore the cave, but the party was forced back by the flooding of the series. On 19 September, a party of fifteen from various clubs, under the leadership of Gerald Platten, who had been one of Roberts' original team, not only crossed the three lakes to the waterfall, but passed a series of rapids, dragging a small canvas boat with them until they came to the fourth lake. This they crossed and followed the dry series beyond, adding a further thousand feet to the 1912 explorations. The following day, a further two thousand feet were explored, when the party returned in case their retreat was cut off by rising lake water.

On 13 October, a fresh team was organized by Platten and three new chambers were discovered. One of these chambers contained a vast area of long straw stalactites. At the end of the month, a further party entered the cave, including myself, the object being to go over the ground which had been discovered and explore avens and side passages which had been passed by in the urge to complete the main explorations.

To approach the river cave, we had to make our way along the river bank. Once inside the cave, we struggled along the river bed, keeping to the side where the water was shallower, until we reached a rope which Platten had thoughtfully fastened across the river to prevent us being swept out of the cave while we made our way across to reach the small, steep tunnel leading up to the dry series. There were of course no wet suits in those days and because our clothes were water laden, the back and knee climb up to the dry series was an ordeal in itself. The water from our clothes ran out, converting the dry footholds into mud, in which it was difficult to get a hold. After considerable struggling, we pulled ourselves into the large dry passage above, now familiar to many visitors as the first passage of the show cave, after the artificial tunnel which is now the entrance to the cave. The passage up which we struggled is pointed out by the guides as one of the first features in the conducted tours.

The 1937 re-exploration of Dan yr Ogof was the major event which established the Tawe or Swansea Valley as a likely area for the discovery of new systems. It was Platten's discoveries which drew the attention of cavers from other well-known caving districts to this area and eventually, after the war, led to the establishment of resident cave clubs in Wales. There is little doubt that much of the impetus for the early explorations in Wales came from the fact that the caving grounds of Gerald Platten and his group on Mendip had proved relatively unproductive for a number of years. A spate of unsuccessful swallet digging on Mendip had encouraged disappointed cave seekers to turn their attention to the almost untouched and not too distant areas of Wales, while a few others pursued their activities in Devon.

In August 1955, some further minor exploration was achieved in the cave with the use of a 'maypole' and in 1965 some diving operations were carried out in the inner recesses of the cave, adding further information to the underwater survey and, in the same year, further passages and chambers were added by 'maypoling' and exploration. The formations found were outstanding, many of the 'straw' stalactities being over ten feet long.

In 1966, a very important extension was added, when two of the smallest members of the South Wales Caving Club managed to negotiate a tight squeeze, followed by other members of the party, who having been assured that there was room beyond the squeeze for their great bulk, managed by excruciating efforts to get into the space beyond and descend a steep aven to a lower series, which proved to be of considerable magnitude with fine formations. They named the first chamber they found Gerald Platten Hall. Finally, after some distance through extensive passages, they came to a lake which they attempted to swim, but decided that a boat would be necessary and they returned to the surface in the early hours of the morning. Within a few days, they returned and crossed the lake to explore larger passages beyond. A halt was made to further exploration at this stage to enable speleobiologists and others to examine cave life etc. in the new series before any further disturbance or the intrusion of any life which might be carried in from the world outside.

Further exploration was done beyond the newly found lake which they called the Green Canal. The South Wales Caving Club were assisted by a party from the Wessex Cave Club later in 1966 and further spectacular discoveries were made, including a chamber with vertical sides and a huge leaning pinnacle of rock, named, of course, the Pinnacle Chamber. The series, because it followed a north–south fault, was named the Great North Road. In 1967, a Yorkshire team found a further series which was named the Far North, terminating in the Great Hall, the furthest point reached. The system obviously goes on, but access through narrow passages prevents heavy equipment being brought through to clear the boulder chokes. I have no doubt that cavers, with their enthusiasm and experience, will one day find a means to overcome the problem.

During the same year, a series of flood passages, known as the Mazeways, were discovered, and in 1968 cavers who camped nearly a week within the cave discovered another series above the Pinnacle Chamber and this series was extended by the University of Leeds Speleological Association.

In 1972, diving in part of the Mazeways discovered about half a mile of dry passages and, almost at the same time. some passages at a higher level were found, including the one now known as Dali's Delight. These passages were later extended, but, like the 1967 series, boulder chokes are at present a barrier to further exploration. Up to the time of writing, Dan yr Ogof consists of known passages totalling more than nine miles.

Dan yr Ogof caves are at present the only commercialized caves in Wales. Besides the formations an outstanding feature of the cave are the huge passages, with the white calcite veins in the dark rock giving the appearance of marble. The first of the formations is not far from the entrance, where an overhanging mass of stalagmite of font-like formation has been given the name of the Frozen Waterfall. It is a type of formation and name which is found in many caves, but is always spectacular. Further along the passage is the white parrot, comparable with the actual bird in appearance and size, although this particular parrot lost part of its tail many years ago.

Further along the passage and outstanding among other formations is the Pincushion, a thick cluster of thin straw stalactites hanging from the roof. Although this group is striking, there are even finer and longer straws in parts of Dan yr Ogof beyond the lakes in passages which the tourist never sees and which are accessible only to cavers. As well as the many formations in the public part of the cave, there are many more extremely fine groups beyond the final point at which the normal visitor begins his return journey.

About 150 yards from the entrance, the Parting of the Ways is reached and flights of steps lead up to higher level passages, taking the visitor past the Alabaster Pillar, a stalagmitic column, the fusion of stalagmite and stalactite between floor and ceiling. The column is nearly six feet high and the junction of stalagmite and stalactite is clearly visible. At its foot is a pool, while many miniature stalactites are to be seen over the surface of the ceiling, near the top of the column. Next we come to the Flitch of Bacon, the name given to a stalagmitic curtain where the translucent red and white bands give the appearance of an enormous rasher of bacon. Further on, the Dagger Chamber is so named because of a pointed stalactite over six feet long, while its corresponding stalagmite is also of slender proportions. Nearby is a group of short whitish stalagmites standing on a stalagmitic flow. At the time when Walt Disney's film *Snow White and the Seven Dwarfs* was current, they were known by the film title, but have since been called the Nuns. Above them is a fine corresponding group of stalactites, some turned at the edges like slender pointed scoops.

At the end of the passage is the impressive 40-feet-high chamber known as The Cauldron, where, from a hole in the roof, hangs a fine wide curtain 18 feet long. From The Cauldron steps lead down to the Bridge Chamber and there, from a natural stone bridge, can be seen in dry weather the boulder-strewn beach leading to the first lake and. under very wet conditions, the edge of the lake itself. From here the tourist returns to the Parting of the Ways, by an alternative way called the Western Passage where will be seen, among other formations, the miniature Elephant whose trunk is really a small stalagmitic column.

For many years the cave of Dan yr Ogof was the only show cave in the complex but Dan yr Ogof caves later incorporated an adjoining second cave. For years following the re-exploration of Dan yr Ogof, cavers had attempted to dig out a small passage in a cliff face between the Show Cave and the Bone Cave (Ogof yr Esgyrn) above it. At first it developed into a miserable little tunnel which they called Tunnel Cave, but when it finally did 'go' the explorers found themselves in a long, lofty corridor with a wide stream running through it. The owners of Dan yr Ogof forced a more convenient way into it, cleared away the boulders and provided a suitable path and lights for visitors. Because of the enormous size of the corridor they called it Cathedral Cave. Cathedral Cave is not a repetition of the older show cave. It is quite different. It has but few formations, but what it lacks in these, it makes up in the grandeur of the passages.

The discovery of such an enormous system as Dan yr Ogof convinced cavers that there were likely to be other extensive systems in the same area. Over the years, the mountains above Dan yr Ogof were searched for likely inlets, either to parts of Dan yr Ogof not then reached or to new caves of similar proportions. The whole area was pockmarked with indications of underground streams, deep swallets or sink holes, like Sink y Giedd and Waen Fignen Felin. It was on the opposite side of the Swansea Valley, however, that the next big breakthrough was to come.

During the war years, some attempt was made to penetrate the stream outlet known as Ffynnon Ddu at Rhongyr Uchaf Farm, but without success. High in the mountain, the River Byfre went to ground at the foot of a limestone bluff of rock and emerged again down in the valley on the banks of the Tawe. Between these two points the Byfre stream flowed underground and it was felt that somewhere there must be a point where an entry could be made into the underground system. Mr Powell of Rhongyr Uchaf drew our attention to an area in the nearby Pant Canol, where he had thought of trying to force a way into the system.

We dug for days and eventually broke into a small spiral-like passage, which took us into a chamber which contained a large pool, waist deep in the centre. We explored every passage we could find but there was no obvious way through into the main system. At Mr Powell's suggestion, we called the small cave Ogof Pant Canol, but it was clear that we would have to seek elsewhere if we were to get into the main course of the Byfre. Later, Messrs Harvey and Nixon broke into the main system at another point nearby and so in 1946 Ogof Ffynnon Ddu was found. Extensive waterways, chambers and dry passages lay open before the explorers, although, like Dan yr Ogof, much more was added in subsequent years. Wales had now two explored large-scale systems, one on each bank of the Tawe, to add to its quota of smaller caves.

Ogof Ffynnon Ddu, today, exceeds Dan yr Ogof in length, but its wet passages and hazards make it a caver's cave and like many caves it does not lend itself to commercialization. as does Dan yr Ogof. It has many fine formations and much spectacular scenery, but it is essentially a scientific cave where studies of all kinds, such as that of helictites, those queer stalactite formations which do not conform to pattern, are carried out. It is a tough cave for exploration, subject to flooding in sectors, and its length and intricacy make it essential not to explore it without the assistance of the South Wales Caving Club who, in any case, hold the keys. In fact, it is always advisable to consult the local caving club before visiting any cave which is not commercialized, whether the cave is gated or not.

Most of the cavers at first concentrated their efforts in the Swansea Valley, because of the two major discoveries there, together with some probing in the known systems near Porth yr Ogof. In the Swansea Valley the Welsh Branch of the Mendip Exploration Society had taken part in the re-exploration of Dan yr Ogof with Gerald Platten's own group and had undertaken the excavation of Ogof yr Esgyrn. The main branch of the Mendip Exploration Society in Bristol ceased to function because of the war but a few members of the Welsh Branch who were stationed in Wales carried on with the excavations and explorations as well as they could during the war years. After the war this branch, with new members, formed a new club, the South Wales Caving Club, and set up their headquarters near Dan yr Ogof. The club's first major discovery was Ogof Ffynnon Ddu. The search for new caves was therefore mostly concentrated in this valley, although Mendip-based clubs made frequent visits to the Little Neath area. Visits were made by the South Wales Caving Club to known caves outside the valley, but with the enormous amount of potential cave area around their headquarters, outside visits were usually of a cursory nature.

Although the first impetus of cave exploration in Wales from 1937 came from 'foreign' intrusion into the principality from the Mendip area, it was not long before Welsh-born recruits began to join the newly formed club. In fact much of its initial success was due to the kindly co-operation of the people of the Swansea Valley and it soon became clear that, as a Welsh-based club, it was to become an integral part of Welsh activities. From the beginning, it was the Club's policy that all newly discovered caves were to carry Welsh and not English names. The English members sometimes found it difficult to pronounce the names of the swallets they were digging. Waen Fignen Felen became known to the two English diggers, Dolphin and Lowe, who spent some years on this dig, as Wiggy Wiggy, a contraction adopted by Welsh and English alike, while Agen Allwed became affectionately known as Aggy Aggy.

Soon after the formation of the South Wales Caving Club, an attempt was made to form a 'federation' of the Club and other independent visiting teams, but it was a short-lived idea. The Club was not always to have the complete monopoly of caving in South Wales and, as was to be expected, the pattern, which had long emerged in other areas began to develop. Industrial concerns and schools began to form independent clubs, and some, such as the Cwmbran Caving Club, have made a fine contribution to Welsh caving. However the strength of the South Wales Caving Club can be said to be largely the result of two factors, its long start free of the rivalries which existed in other areas and the broad basis of its membership—local people with their knowledge of the area and cavers from all trades and professions, drawn from other caving areas because of the almost untouched possibilities of new cave discoveries. Like other leading caving clubs in Britain, the club is well known outside its own territory and indeed outside the British Isles, and was particularly honoured by Yugoslavia for the recovery of the bones of certain of that country's war heroes from a deep Yugoslavian pothole.

The third major system to be discovered in Wales was not until three years after the discovery of Ogof Ffynnon Ddu. This time it was further afield near Llangattock in the Vale of Usk and the discovery was made by a party of scouts. The cave Agen Allwed (the Key Hole) was entered from a slot in the cliff face overlooking Llangattock. The difficulty of the

cave depends on one's girth, as most of the way to the main
stream passage is small and involves crawling and squeezing
with some nasty turns and a particularly awkward sector
known as Sally's Alley, but once past these, the stream
passage is much easier going. In 1949, a further extension to
the cave was opened by the Hereford Caving Club.

Although the system is extensive, covering, with its
tributary passages, some fourteen miles, Ogof Ffynnon Ddu
still remains the largest and deepest system in Britain, at over
twenty miles of passages and a depth of 850 feet. It was in
Agen Allwed in June 1974 that Roger Solari, one of the most
talented cave divers whose achievements were known in
several cave areas including Ireland, was killed during diving
operations. Another cave diver, Paul Esser, who had over
eighty dives to his credit, was killed in Porth yr Ogof in
February 1971, a cave which had already claimed the lives of
cavers while swimming through the cave in June 1957 and
October 1970.

A number of other caves, smaller than the three major
systems, have been opened up in South Wales as the result of
caving activities. Paul Dolphin and Colin Lowe, in the early
days of the South Wales Caving Club, spent several years
digging out a swallet at the edge of a boggy area known as
Waen Fignen Felen on the mountain above Dan yr Ogof. It
was one of the most frustrating digs known. Armed with
buckets, spades, picks and ropes, after a gruelling walk up the
mountain, they would dig out part of the entrance to find the
next day that flood water had again filled it in. However they
decided to beat Nature to it and broke their way into a small
chamber, only to find the next day that the roof had
collapsed. They were in some way recompensed for their
years of abortive digging by their success at Pwll Dwfn, best
described as a 'pot' rather than a cave, as it consists of five
descents, each after the other, requiring no less than five
caving ladders, totalling 215 feet in all, the lowest point being
310 feet below the surface. Pots or potholes, deep vertical
descents direct from the surface, are more common in
Yorkshire and Derbyshire than in caves in the southern part
of Britain, but they occur occasionally. This was not the first
pothole to be found in Wales: Pant Mawr, a large pot on the

mountains between the Swansea Valley and Upper Neath
Valley, has always been known to cavers, as it has always
been open, although it is now protected by a gate.

Another cave, which like Porth yr Ogof has always been
open, is away from the main cave areas and is at Trapp, near
Carreg Cennen Castle, Dyfed. It is Llygad Llwchwr (the Eye
of Loughor). The River Loughor flows through it and legend
says that its source is in the high mountain lake of Llyn y Fan
Fach (the Little Mountain Lake), some six miles to the
north-east. The Loughor finally flows out to the sea near the
town of that name. It is one of the sporting caves, favoured by
cavers from early days, but is now controlled by the Water
Authority. A Mr H.T. Jenkins wrote in his diary in 1841 that
he and Mr Peter Jenkins visited the cave on returning from a
fair. Mr Peter Jenkins took his clarinet and the diarist his
pistol and they spent three and a half hours in the cave.
Apparently the clarinet and the pistol were taken to test their
reverberatory effects, for, according to a later diary entry in
August 1843, Mr H.T. Jenkins again visited the cave and
stated that the pistol report rivalled the sound of thunder.
Fortunately the noise didn't have any undue effect on loose
rocks. Mr Jenkins must have been fond of making noise in
caves, because he went again to Llygad Llwchwr the
following month, when he described how he struck a stalagmite formation with a hammer and it gave out a sound like a
bell. We must be thankful that even in those days not all
explorers went about hitting formations with hammers.

The cave was entered by a slot above the river exit from
which a bit of crawling along a zigzag passage and some
short ladder work led to the course of the underground river.
You can get round the deep pools on narrow rock platforms.
On one occasion, I began to overbalance on one of these,
saying to my companion in a quiet voice, so as not to disturb
his balance, "I'm falling in." "What did you say?" he asked,
by which time I was clearly about to take a cold plunge.
Llygad Llwchwr is an intricate cave on more than one level
and once when two of my friends were exploring the cave,
their lights failed and their matches were too damp to light
the emergency candles. It is a serious caving offence not to
have a reliable standby light and as they had both been

recently appointed members of the new Cave Rescue Organization, an organization still administered by the South Wales Caving Club, they could not bear the thought of being rescued by their own team. They ignored the rule to stay put and inch by inch, keeping in contact by touch and speech, they slowly made their way in the darkness through the intricate system, which fortunately they knew very well indeed, until at last they reached the entrance. It is a rule of all reliable clubs that members should leave a note of their caving plans and the time by which they should have returned. If they fail to report by this time, the Rescue Organization is called out. My friends got back to the club headquarters less than half an hour before the crucial time. They had their legs pulled for being late, but it was a long time before they told the real reason for their lateness.

One of the clubs which has been an active visitor for many years in the Upper Neath area round Porth yr Ogof and Pwll y Rhyd (Pool of the Ford) is the Bristol University Spelaeological Society in conjunction with the Cave Diving Group. They have made this their particular area for cave exploration and the study of underground hydrology. As a result, they have done outstanding work, both at Porth yr Ogof and various small passages which abound in the Little Neath area, proving that these are part of an intricate underground system.

An area which has come under special attention by the Welsh Clubs and the Hereford Caving Club and which, as a result, produced the major system of Agen Allwed is Llangattock, near Crickhowell. Above Llangattock is a long impressive scarp, which has been quarried in places. Along the foot of the scarp, at some height above the valley bottom, is the track of a former tramway, a favourite walk giving views not only of Llangattock below, but also of the well-known heights of the district, such as Pen Cerrig Calch (Limestone Top), Table Mountain, and the Sugar Loaf at Abergavenny, with Skirrid Fawr (known as the Holy Mountain) peeping over its shoulder. From this track, several caves can be reached in the rock face. In a quarry in the face is Eglwys Faen (Stone Church), a cave with three entrances. The two small entrances are just above the path, but the larger entrance is about 30 feet up, an easily accessible climb. The cave has an outlet to the moorland above by a 30-foot aven, and in this sense can be said to have four entrances. The large entrance leads into the main chamber, connected by a passageway to the two holes by the path. It is not a large cave, but if you are prepared to crawl, it contains some exciting passages. At the far and western end of the old tramway is Ogof Gam (the Crooked Cave), a long rift which leads into Agen Allwed, the key-shaped entrance of which is a few yards on, but at a slightly higher level. Like most extensive systems, the whole of Agen Allwed was not found in one day. Extensions to the system have been found at different times and by different teams, not all from the same club.

Another area of South Wales with caves, but with comparatively small caves, is the Gower Peninsula. It has long been famous for its numerous bone caves, such as Bacon Hole, Minchin Hole, Bosco's Den and the well-known Pavil-and Cave, in the cliffs of the southern shore. Such inland cave systems as exist are small. Although at Llethrid the system extends to several hundred feet, the passages are small and wet. The Guzzle Hole, near the spot where a stream emerges in the Bishopston Valley, has been a curiosity from time immemorial. The underground water can be heard 'guzzling' away quite plainly from the surface. The more ancient of the Gower sea caves are above sea level, being formed at a time when the sea was at a higher level than today. These may be formed in part by inland stream action, but none penetrates into the cliffs for any distance. The largest of them is Minchin Hole, Penard, and that is a single chamber. They are difficult to reach and it is easy to get cut off by rising tides. Goat Hole, better known as Paviland Cave, the home of the famous British Stone Age skeleton known as the 'Red Lady', is accessible only at the very lowest of tides, except by way of a rather hair-raising ledge. I once visited Paviland as a member of an international congress and, on the return journey, had to piggy-back a couple of the elder members across the rising tide, between the sloping rock spur leading up to the cave and the shore.

While working at Minchin Hole, I often watched the sea

foaming up the gully to the cave entrance and, at high tides, washing into the lip of the cave. As our work there lasted for many years, we got to know the routes up the cliff face, but often had a drenching when heavy seas hit a large rock in the gully, throwing spray up the cliff. The sea gully was a convenient way of getting rid of surplus cave debris for it took only a year or so for some of the largest pieces of rock removed from the cave deposits to break up under the incessant pounding of the tides. I say 'surplus' because parts of the cave deposits were kept for analysis and identification. The removal of excavated debris is always a problem on excavation sites, but at Minchin Hole there were no problems.

In cave areas there are usually one or two isolated caves at some distance from the general cave groupings. In Gower there is Cat Hole, close to the restored New Stone Age burial chamber of Parc le Breos. A path leads from Park Mill on the main road from Swansea to Rhossili to a pleasant valley with a wide greensward between wooded slopes. In the middle of the valley are the remains of the burial chamber, while a little beyond, a narrow path leads to the right into the wood to the base of a slope. At the top of the steep incline can be seen Cat Hole, the entrance partly masked by the old debris mound from archaeological excavations once carried out there. It rather reminds one of Thor's Cave in the Manifold Valley, Staffordshire, although Thor's Cave is higher and more impressive. Cat Hole is a typical bone cave and not an exploratory system, but well worth visiting for the walk alone.

Another isolated cave is Lesser Garth Cave at Tongwynlais, near Cardiff, also surrounded by woodland. Although it is an archaeological site and a small cave, it provides a certain amount of limited exploration. Hoyle's Mouth at Tenby and Coygan Cave, Laugharne, are both archaeological sites, but very limited caves.

North of Merthyr Tydfil, there are two fairly extensive systems, although liable to flooding, Ogof y Ci (Cave of the Dog) and Ogof Rhyd Sych (Cave of the Dry Ford).

The South Wales Caving Club is particularly fortunate in their headquarters. Just after the war, they settled into two cottages near Dan yr Ogof caves, but these were very dilapidated and in the 1950s they moved to a street of disused quarrymen's dwellings in the quarry village of Penwyllt (Wild Top), high up on the mountains on the other side of the Tawe valley from the show caves. The club occupies the main terrace of houses known as Powell Street, the cottages now converted to provide changing and drying rooms, clubrooms, dining rooms, dormitories and showers. There is even a workshop, fitted with engineering plant, lathes and battery-charging facilities for cavers' lamps, and a research laboratory. They make nearly anything there, including metal pliable ladders for cave descents, metal shoring for new shafts, and 'maypoles' for reaching inaccessible places. The facilities there bear out all I have said about joining and caving with a good club.

TEN

Caves of Mendip and Devon

THE MENDIP CAVE AREA is compact, and although the range extends about forty miles from Frome to the sea at Weston-super-Mare, the principal caves are in the central part, mostly round Priddy, Charterhouse and Cheddar, although two large systems are in the eastern part of the area.

The most famous feature of Mendip is Cheddar Gorge. Viewed from across the flat land to the south, it looks like a huge zigzag crack into the hills. It can be called the southern gateway to Mendip for the winding road leads you up into the heart of the hills. At the foot of the gorge are the equally famous caves of Cheddar. Although another well-known Mendip cave, Wookey Hole, was open from time immemorial, as far as is known, the Cheddar caves were not discovered until the last century. However, there appears to be some mystery about Cheddar caves, for in 1125–30, Henry of Huntington wrote of "Cheder Hole where is a cavity under the earth, which, though many have often entered—they could yet never come to the end".

In Holinshed's *Description of Britaine* of 1568 is a further reference to Cheddar Hole, "whereinto many men have entered and walked verrie farre". As both accounts refer to internal streams or rivers, it appears that people must have been able to penetrate much further than they can today. It is possible that this cave may have been entered through Cooper's Hole, the large silted hole adjoining the road above the commercialized caves.

In 1837, Edward Cox, who ran the Cliff Hotel and the mill adjoining it, removed some rock at the foot of the cliffs to accommodate his carts and provide a carriage space for the hotel residents. While doing this he found the entrance of a cave which he commercialized as Cox's Cave.

In 1880, Richard Cox Gough and his sons, relatives of the

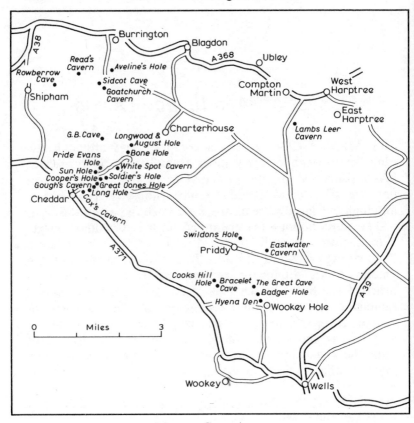

MENDIP CAVE AREA

Cox family, who had been digging various crevices and holes in the Gorge, turned their attention to a partly blocked cave behind a cottage occupied by Jack and Nancy Beaumont. The Beaumonts had been allowing members of the public to view it for a few pence each. R.C. Gough apparently acquired an interest in the cave for he improved it, opening it to the public on a larger scale. This became known as Gough's Old Cave and in 1893 he and his sons forced their way into the high rift which contains the gours, known as the 'fonts'. To this they added more passage by blasting. It was not util 1898 that they found King Solomon's Temple and St Paul's, while digging for further extensions.

Many thousands of visitors visit Cheddar each year for the extraordinary beauty of the stalagmitic formations, and many look into the museum to see the skeleton of an Old Stone Age man, known as Cheddar Man. This young man died when he was twenty-three years of age and was buried in deposits which were found to contain one of the finest collections of flint blades and the only two or three true Magdalenian relics found in this country together with a harpoon from Aveline's Hole, Burrington Coombe. In Cheddar Gorge are a number of small caves and rock shelters, some of them of considerable archaeological importance.

The other well-known show cave of Mendip is Wookey Hole, not far from the city of Wells. Wookey Hole is the name of the cave and the village. The cave has also given its name to the village of Wookey two miles away, which indicates that the name is older than both settlements. It is said to be of Celtic origin, derived from 'Oky' which in turn came from 'Ogo'. The Welsh word for cave is *ogof*.

The cave always seems to have been open to visitors and there are many accounts of early visits. Alexander Pope, the Poet Laureate of his day, famous for such poems as 'Essay on Man' and 'Rape of the Lock', knew Wookey Hole, for he had some of its finest formations shot down from the Hall of Wookey and taken to Twickenham to adorn his private grotto there. There were no cave conservationists in his day. Had there been, I am sure one of them would have written a satirical poem on the 'Rape of Wookey Hole'.

The legend of the Witch of Wookey has already been told in the chapter on cave legends. Mr H.E. Balch, the 'grand old man of Mendip', who excavated in the Great Cave in the early part of this century, was convinced that he had found the bones of the witch, her two goats, and her milking pail. Most remarkable of all, near the bones was a ball of stalagmite. Was this the witch's crystal ball? Strangely enough, a map attached to the ancient work known as the *Polyolbion* depicts Wookey Hole with an old lady at the entrance, who appears to be holding what could be interpreted as a crystal ball. These relics, together with other finds from Wookey Hole, may be seen in the museum at Wells. The bones are a mixed lot, as other human bones were in the vicinity and seem to have been mixed with those thought by Balch to be the witch. The sex of the old goatherd depends upon which bone you are looking at.

The removal of many of the finds to Wells Museum meant that there was not a great deal of material left when the Wookey Hole caves museum was set up some years ago. However, the Cave Diving Group during their operations at Wookey Hole recovered from the river bed a good number of skulls, bones and some interesting pots of a period when Britain was under the Roman occupation. These, together with other relics including those found in the Fourth chamber in more recent years, make an interesting museum at the cave. There is the skeleton of an arthritic old lady in a glass-covered coffin-like case on the floor. She was found by accident. I was asked by one of the caving clubs who operate in the area to examine an animal bone they had exposed on the floor of a rift close by. As I was descending the narrow rift, the wall opposite suddenly collapsed. When the dust cleared away, I found myself looking at a hole in which I could see a couple of skeleton feet and the leg bones beyond. A couple of days' digging from the surface took us down to the skeleton, which is probably Iron Age or Romano–British. We never found the animal bone.

To visit Wookey Hole caves, you follow a very pleasant path above a small valley on one side and pleasant laid-out gardens on the other. Along the bottom of the valley can be seen the River Axe and a short canal leading to the paper works. The river flows out of a wide natural arch and tumbles down over the supporting wall of a dam. I remember standing on the sill of the dam, waiting for the divers to bring out a Roman lead ewer which had been seen lying, not far within the arch of the cave, in a depression in the river bed.

Across the valley, hidden by trees, is the famous Hyena Den, a single-chambered bone cave, partly open to the sky. It was here that Boyd Dawkins carried out his famous excavations which revealed the hyaena bones which gave the cave its name.

On a higher level can be seen the debris heap of Badger

Hole, dug more recently by H.E. Balch, Boyd Dawkins' pupil and successor in the Wookey Hole area. In front of the debris heap is the stone wall from which we would occasionally climb down to the Hyaena Den below, or pass behind to reach the Badger Hole, where Mr Balch was seated at the edge of a large platform of planks on which bucket after bucket of cave earth was heaped for him to sort.

Back to the Great Cave, we pass through the iron door and there in front of us, a huge rock is supported by an artificial pillar. It was near this rock that Balch found the stalagmite ball and the bones of the goatherd. When Balch started his excavations the whole area below the rock, including the space occupied by the pillar, was full of cave deposits containing quantities of Romano–British material. The only way into the interior of the cave was by squeezing round the free side of the huge rock and this was the only way the early explorers and travellers could enter the cave.

The local people who acted as guides would illuminate the way with flaming torches. Beyond, the passage floor was strewn with boulders and the descent down Hell's Ladder and across the floors of the chambers beyond must have been hard going. No wonder the flickering torches of the guides throwing eerie shadows here and there conjured up a vision of Hell, even of the dark waters of the Styx, while the silhouette of the stone witch herself, the 'Sorceress', must have stood out in sharp profile as a guide poured spirit on the walls of the cave and ignited it. Today, with the strong cave lights, you can see all the features of the chamber in one colourful panorama. The first underground television documentary ever to be transmitted alive was from this chamber.

The second public chamber, the Hall of Wookey, is almost a huge passageway between the first chamber, the Witch's Kitchen, and the third chamber, the Witch's Parlour. Because of its narrowness, its great height appears to be even higher and it is little wonder that the stalactites which once hung from the ceiling had to be shot down for Pope's Grotto. It is hard to imagine that much remained of them after they had made their quick descent.

Between the Witch's Kitchen and the Witch's Parlour, the River Axe flows through a submerged passage, and it was in this passage that most of the relics, bones and skulls were recovered by divers from the bed of the river. In fact, this submerged passage leading off from the Witch's Kitchen was known to the divers as the Skullery!

At the end of the Hall of Wookey, you pass into the third chamber or the Witch's Parlour, a very broad, sandy-floored chamber, not high, but with a marvellous single-span roof which produces startling acoustics. At the far end of the chamber is the arch of the submerged passage leading to chamber four and beyond. In 1947 the Cave Diving Group began their operations to explore the unknown inner reaches of the river. Over the years, many new passages and chambers have been discovered, some totally submerged and some dry. Until recently these chambers could only be seen by divers, but now the public can see three of them, as the owners of the cave have completed a dry artificial tunnel where views of these areas, previously only reached by divers from chamber three, can be seen. This tunnel is now the exit from the cave and brings the public out into the open in the resurgence valley, not far from the Hyaena Den.

When H.E. Balch first explored the Mendip caves, apart from Lamb Leer and the show caves at Cheddar and Wookey Hole, there were very few known caves. It was Balch and his friends who, after Martel's descent of Gaping Gill, started the impetus at the beginning of the century which led to the popularity of Mendip caving. This new interest in caves was due, in part. to the many illustrated talks he and his photographer Savory gave following their major discoveries and the books which were written by him and his friend, Dr Baker.

Balch has been called the pioneer of Mendip caving, but this is only true in the search for new caves. There are many old accounts by adventurous travellers of caves they visited, but what Balch and his contemporaries in other regions did was to search and dig for new caves. The first dig for a new system by Balch was in 1901, when two of his men broke into Swildon's Hole. A few days later they were joined by Balch and reached the top of the 40-foot waterfall, but it was years before that was passed, as after the initial discovery the owner closed the cave and, although clandestine visits were made, Dr Baker having found a key to fit the padlock, full-scale excursions were not possible for some time.

It was the owner's action at Swildon's Hole which was responsible for the discovery of Eastwater in 1902, as, barred from exploration in Swildon's, the team turned their attention to another swallet in the same district. They temporarily dammed the stream and began digging out the swallet. At one point when investigating a fissure in the cliff face, a rock fell behind Balch and Willcox, trapping them in the fissure, but it was quickly broken up by the rest of the party. The swallet-digging took some days, three more days than were necessary, as one night a cow broke its neck in the swallet and it took that time to remove the carcase. The first obstruction within the cave was the great ruckle of boulders through which a route had to be found to reach the solid cave passage.

Since that first breakthrough, a great deal of further exploration was made by Balch and his party and others in the following years. In 1910 objections to visiting Swildon's Hole were lifted. At some time before 1914, the first descent of the waterfall was made by Ward, who was lowered down the drop by Barnes and a lad they had brought with them. They waited an hour but nothing happened. Ward was in the pool in a weak condition and unable to light his candle. He was pulled up, an arduous task for a man and a boy. He seemed to recover when he reached the top, and left the cave cheerfully. A few days later Balch called on him to find him gravely ill after his adventure.

In 1914, Dr Baker, with two of his Yorkshire friends, Chandler and Roberts, succeeded in passing the fall, using rope ladders. In spite of a bad foot, Roberts made the first descent and like Ward had trouble in striking his matches on the damp matchbox. Fortunately he had waxed his matches and, by scraping one of the damp walls reasonably dry, he got his candle alight. He then went on to the second pitch and returned for Baker. Together they reached the twin pools, but Roberts' foot was getting worse and they had insufficient tackle. They had to return. The following week, Roberts' foot was operated on for blood poisoning.

In 1919, Roberts intended to pass the fall again with Ellis, but he dropped his spectacles down the pitch and Ellis was not a climber. Although Roberts apparently made the descent, they felt the odds were against them. A further attempt was made in 1920, but the water was so great that they turned back.

On 23 July 1921, Baker, Chandler and two others got over the fall again and past the 20-foot fall for some distance, where they erected a cairn. A week later, an advance party started to clear the way at the head of the 40-foot fall in readiness for a big onslaught by Balch and his team on 1 August, with the assistance of four members from the University Society. There had been two seasons of drought and the water going over the fall was negligible, but it took two hours to complete the clearance started the previous Saturday to enable the large quantity of tackle to be taken through. In the afternoon, they had not only passed both waterfalls and reached Baker's cairn, but explored new ground beyond, including the exceptionally beautiful Barnes' Loop.

An all-night visit was made to the cave in November by a party of fourteen, led by Balch for the purpose of photographing Barnes' Loop, when further passages and outstanding formations were found by members of the University Society. At one point, their way was barred by formations and rather than destroy any of them, they were left untouched until 1945, when a Royal Air Force Club decided to break through. They found a finely decorated grotto, but further progress was barred by stalagmite. In 1953 another party managed to remove enough of it to pass through. Beyond was a large, long passage with more beautiful formations. The passage ended in a rather muddy and small tunnel, leading to a network of narrow ways, ending in a mud sump. This network they called Damascus and the passage, decorated with stalactites leading to it, St Paul's Grotto, as it was first found on that saint's day. The sump was eventually passed in 1955 and the series of rifts beyond was called Paradise Regained.

The main stream passage ended in a sump which, for many years, barred further progress. After the war, with the advent of the techniques of sump swimming and diving, several sumps were passed and explorations added a great deal to the known system.

Although Baker's and Balch's trips over the waterfalls were by independent teams, there was some liaison between the co-authors of *Netherworld of Mendip.* Each kept the other briefed about their activities. In fact, some of the tackle used by Balch on 1 August 1921 was lent by Baker and there is little doubt that he would have lent more, had he not required tackle for another project on the same weekend. Although Baker may have stolen a march on Balch by passing the fall a week before Balch's expedition, Baker already held the distinction of having crossed it in 1914, and Ward had descended before him. Any annoyance expressed by Balch's followers about Baker's 1921 descent was not shared by Balch, who in a written statement on 27 July 1921 said he did not mind Baker going down at all, and indeed was rather glad about it. Balch probably had far less resentment against his old friend Baker, who occasionally came in from the north, than he had against the fairly new University of Bristol society. He was inclined to think that this society intruded on the territory of his own well-established society, the Mendip Nature Research Committee. An arrangement was made between the two societies that the "Bristol men", as Balch called them, would operate in the Burrington Combe area, but any central Mendip trips would be by his own society's invitation only. This arrangement operated for many years.

From the letters Balch wrote just before his descent, it is clear that all went according to plan, with only one omission. His theory was that the stream that formed the cascade originally descended by a continuous slope from the end of the wet way to the floor of the waterfall chamber and that a great accumulation of boulders had choked the old channel, causing the stream to flow over this boulder choke to form the waterfall. On 26 July, he wrote to his friend Savory that he intended to shake up the boulders in the floor at the entrance of the water rift leading to the fall, or the Pithead as Balch called it, with an explosive charge. This he hoped would allow the stream to seep down by its original course and so dry up the waterfall for the descent. He suggested that a dam might be temporarily erected at the end of the wet way while this was being done. As it happened, the explosives taken for this purpose were never used, as the fall was almost dry, but it is interesting that during 1968, excessive floods caused the fall to collapse and the resultant slope now conforms more to Balch's idea of the original channel.

Although Balch spent many years in expeditions to the Mendip caves, often exploring new ground, he had only one major accident. That was in Lamb Leer, Harptree, Mendip, when a rope broke and he fell 60 feet into the Great Chamber. Although he did not remember it, he must have caught hold of the cotton rope which had been placed for steadying his descent, and so broke his fall. His fingers were cut to the bone. He recollected that when he recovered consciousness on the floor of the rocky chamber, he saw the light of the hauling party far above, and for a moment he thought he was in another world. Balch's description of the fall in his book *Mendip—its Swallets, Caves and Rock Shelters* is so vivid that there is little object in repeating it.

Lamb Leer itself has an interesting history. It was found during mining activities in the seventeenth century and a description of it was recorded in the *Philosophical Transactions* of 1700. The entrance was obliterated and its actual position lost for 150 years, although the memory of the cave was passed on to further generations. When, about 1879, mining activities were resumed in the area, the Manager, Mr Nicholls, arranged for a search to be made for the lost cave. The land all around Lamb Leer is so pitted with old workings that it was difficult to locate. It took thirty-seven borings before they found it and the first account of the rediscovery appeared in the *Times and Mirror* on 26 June 1880. The miners erected a windlass and a cantilevered platform where the entrance passage broke into the top of the main chamber, 65 feet from the floor. A descent was made by the mining agent, James McMurtrie, his son, Mr Sopwith and Mr Wynne, and a survey of the cave was made.

In the cave are the initials of Thomas Willcox, once manager of Priddy Lead Works, with the date 1894. Balch decided to use the inscription as a yardstick for the growth of stalagmite as had been done with the Hedges inscription at

Kent's Cavern, Torquay, but for reasons explained elsewhere in this book, the method is no longer accepted.

The winch installed in 1880 has since been removed and is now an historic relic in the museum at Wells. When I visited the cave some years ago, the descent to the floor of the Main Chamber was by a box on a sloping aerial cable. The only snag was that it would occasionally jam about halfway along the cable, and today descents are usually made by pliable ladder.

A new chamber was found in the cave in December 1970 by the Speleo Rhal Caving Club of Southampton when digging in a 3-foot-high tunnel called Promise Passage which is reached by such sections of the cave known as Misery Crawl, Corkscrew and Agony Crawl, which all speak for themselves. The chamber is 60 feet long, 25 feet high and 20 feet broad and has been called December Chamber. Some years previously, Professor Palmer carried out a geophysical survey on the land above Lamb Leer and his results indicated an additional chamber to those already known. This unseen chamber was named Palmer's Chamber. It was at first believed that this chamber had been found by the Southampton team, but on comparison with the survey, it was found that this was not so and it would seem that there is still an undiscovered chamber in Lamb Leer.

The next largest cave to be found was G.B. at Charterhouse, named after its discoverers Goddard and Barker, in 1939, as the result of swallet digging. On the first day of exploration, the two men found themselves in a descending narrow passage, leading into a larger rift, where they found the First Grotto and a larger chamber. Both contained formations, including helictites. Opening out of the larger chamber and at a higher level was the Upper Grotto. They then went on through a long small tunnel, so full of mud and water that they named it the Ooze, but they were unaware that it would lead them nowhere, and they had to return as they had not sufficient lighting to go on. Further exploration revealed an alternative series of narrow passages and the enormous and impressive main chamber for which the cave is well known. One of the tightest parts of the cave is the Devil's Elbow and a number of cavers have exciting stories of their efforts to struggle through it. I must admit that it is one of the most difficult bends that I have encountered, but an alternative route is now used. New discoveries were subsequently made in the cave, and in 1966 a further extension of 700 feet was added.

Close to G.B. Cave is the combined system of Longwood Swallet and August Hole. Longwood Swallet is one of a series of swallets in the area and was cleared by the Sidcot School Speleological Society in 1944. The cave is entered by a shaft. At the bottom, you squeeze through into a low bedding plane which many cavers find the most difficult part of the cave. A 10-foot drop leads into the Great Rift Chamber from which a further descent and a drop of 33 feet brings you into the Great Chamber. The final exploration of the Great Chamber and the passages leading off was the work of two other caving clubs as well as the Sidcot members. There was every indication that the final exploration had not been reached, and in 1947 it was noticed that a large amount of sediment had disappeared down a rift. Boulders were removed from it, opening up the extended system known as August Hole.

In the Longwood Valley in 1971, after three years' work, was found Rhino Rift, which contains three pitches of 100 feet, 50 feet and 70 feet, separated by steep inclines, giving a total depth of about 400 feet. Because of the engineering techniques used, it has been described as "one of the most mechanized digs on Mendip".

An extensive system was also found at Manor Farm in the same area and this was one of the most tricky swallet digs undertaken. The first attempt to open the system was made in 1947 by the Aces Caving Club, without success. Further attempts were made by various other groups and about 1952 the Aces renewed their efforts again, but without success. In 1955, the University Spelaeological Society decided to make an attempt, but like the previous groups had to abandon work because of the unstable condition of the swallet. In 1972, a combined party from several clubs, with the university team, continued the work and an entry was made into one of the chambers, known as Penthouse Chamber, and after a large amount of blasting, the cave was finally cleared in September 1973, revealing an extensive system.

For some years, the Bristol Exploration Society worked on an exceptionally difficult swallet near St Cuthbert's old resmelting works at Priddy, but they were eventually rewarded with the discovery of one of the most interesting caves on Mendip. The system is very extensive and difficult in places, but climbing aids have been installed to make these safer.

East Mendip is not so rich in caves as the Charterhouse and Priddy areas, but this area does have some important caves. Stoke Lane Slocker was entered by an exceptionally wet passage, but was gradually extended by passing several sumps. Further sections known as Stoke Lane V and VI were found in 1965.

Thrupe Swallet was first attacked in 1936, but in recent years the Wessex Cave Club have found quite an extensive system.

Fairy Cave, once an easy and pretty cave, has unfortunately been damaged by extensive quarrying, as have some other caves. In the same quarry is the much more extensive Shatter Cave, which would have suffered the same fate had not conservation action been taken. It consists of a series of passages and chambers, and although some of its formations have disappeared, it still has a number of outstanding stalagmites. It was named Shatter Cave because of the shattered condition of the rock at the entrance.

On the northern side of Mendip is Burrington Combe, famous for the 'Rock of Ages', where the Rev. Toplady was inspired by the shelter afforded by the rock cleft. Aveline's Hole, a little higher up the Combe, would have given him better shelter, but it was not discovered until about thirty-five years after the hymn had been written. The discovery in 1797 was accidental, when two men were trying to dig out a rabbit. The large opening descends on an incline into the cliff, ending in a depression at its far end. It is essentially a bone cave, as a number of Old Stone Age human skeletons and artifacts were found there. It was named Aveline's Hole after his teacher and friend by Boyd Dawkins, one of the several excavators.

There is no doubt that the cave connects with the shaft of

Plumley's Den in the adjoining quarry as sounds made in one can be heard in the other. The shaft was named after Plumley, who was lowered down when it was discovered in 1875, but was killed when being pulled up again, apparently breaking his back against an overhang. The shaft was partly filled with stones and a tree soon after the fatality. Subsequently the mouth of the shaft was covered by a slab. At one time, during the first decade of the century, the Lord of the Manor, Mr Gibson, intended to clear the shaft and explore beyond.

He also gated Goatchurch Cave in the lower Twin Brook Valley, a small tributary ravine higher up the combe, to prevent villagers, who had been taking specimens of calcite, from further damaging the cave. It is nice to think that cave conservation had its devotees even before 1907. However Mr Gibson must have had in his mind the possibility of opening the cave as a show cave, for steps were cut in the sloping floor of the entrance chamber and the old handrail supports still remain, although the gate disappeared some years ago. The fact that the public were admitted on certain days of the week or could view by appointment seems to bear this out. The show caves at Cheddar had only been open to the public for a few years and no doubt were an incentive to try and commercialize Goatchurch Cave. The project was short lived, but the entrance chamber of Goatchurch still remains the easiest uncommercialized cave to be seen by non-caving but adventurous people. These stop at the end of the sloping entrance chamber, where a muddy cleft leads down from boulder to boulder, the Giant's Steps.

Goatchurch is a good cave for training the beginner, as it is a fairly dry and safe cave if you don't rummage about in the water chamber. The difficulties increase as you go along, so the beginner can always stop when he has had enough. Now it is so popular for this purpose that I have known queues at the entrance.

One of the youngest of the cave areas is Devon, where caving activities were started by a group of former members from Mendip. The area contains comparatively small limestone outcrops. That is not to say that Devon had no cave tradition. It had already made a name by the contributions of

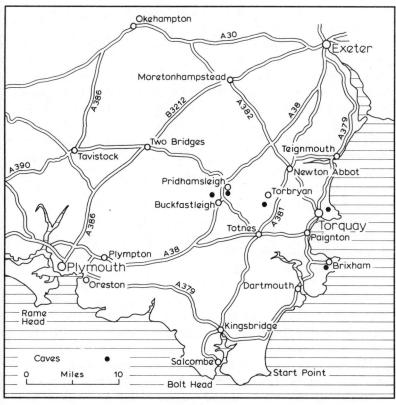

DEVON CAVES

MacEnery and Pengelly at Kent's Cavern, Torquay, and at Brixham Cave. Although these caves are chiefly famous for their prehistoric material, the system at Kent's Cavern is reasonably large. Devon has an extensive cave system at Buckfastleigh and another at Pridhamsleigh and there is the usual collection of small caves, including important prehistoric sites, but if any area is a centre of cave research, Devon certainly is. The original work at Kent's Cavern was a contributory factor in dispersing the bigotry and prejudice which surrounded our first understanding of man and his background in Stone Age times, and research into the formation of stalagmite stimulated interest in these deposits and in caves generally.

It was not because of the importance of the work carried out in Kent's Cavern and Brixham Cave that Devon was chosen for the establishment at Buckfastleigh of the William Pengelly Cave Research Centre in 1962, although in itself that might have been a fitting reason. Its establishment was the outcome of the fears of the Devon Speleological Society that the caves in Higher Kiln Quarry, Buckfastleigh, which they had explored and carried out research in for many years, might be threatened by an impending change of ownership. Cavers are well aware of the threat to caves by quarrying and dumping of rubbish, and in a tourist area like Buckfastleigh the caves might well have been acquired as show caves. Show caves have their merits in presenting to the public a world which they would not otherwise experience, but it is sometimes difficult to carry out certain forms of research in them. The Devon Speleological Society had been carrying out research on bats, animals highly susceptible to public intrusion. The quarry was on a list of properties to be designated by the Nature Conservancy as a site of Special Scientific Interest, but there was no guarantee that under new ownership, the Society would have access to these caves or be able to continue with their research.

The fears of the cavers were brought to the attention of the Society for the Promotion of Nature Reserves and after considering the work already carried out at Higher Kiln Quarry, the Society purchased the site at public auction and leased it to the Devon Trust for Nature Conservation with the old buildings the quarry contained. To administer the centre, a new body was formed, the Association of the William Pengelly Cave Research Centre, which included members of the Society for the Promotion of Nature Reserves, the Devon Trust for Nature Conservation and the Devon Speleological Society.

The purpose of the William Pengelly Cave Research Centre was not only to carry out research, but to promote education in various aspects of caves and cave conservation. The quarry contains several cave entrances, although they all form part of the same extensive system. Most of these caves are used to demonstrate a different aspect of caves and caving. The largest of the entrances is Reed's Cave, which contains very attractive and interesting formations and is a sporting cave. A few yards to the left is Disappointment Cave, perhaps the least interesting of all the cave entrances. Then comes Rift Cave which is used for bat research because of its colony of Greater Horseshoe bats and a minute blind and colourless kind of shrimp, known as *Niphargus glenniei*, peculiar to certain Devon caves.

To the left of Rift Cave is Joint Mitnor Cave, a very important member of the group because it produced a large collection of animal bones of the last interglacial period. This collection can be seen in the Torquay Museum. The bones include those of hippopotamus, soft-nosed rhinoceros, and the straight-tusked elephant which was the forerunner of the mammoth of the last Ice Age. The Centre has cleared fallen boulders from the cave and has made a cutting showing bones still in position, as a demonstration of bone cave excavation. There was a danger that research workers and visitors, entering Rift Cave to see *Niphargus glenniei*, might disturb the bat colony, and so specimens of the shrimp-like creature were transferred to a special area prepared in Joint Mitnor Cave where they seem to be flourishing very well. To the left of Joint Mitnor Cave is Spider's Hole, so called because *Meta menardi*, the cave spider, is well represented there. To the far right of the line of caves is Partition Cave.

Close to Higher Kiln Quarry is Baker's Pit, an extensive sporting system, but further afield at Pridhamsleigh is Pridhamsleigh Cave, which is also one of the most extensive caves in Devon. At Torbryan, near Newton Abbot, is a string of ten small caves which produced archaeological finds, some of which are in Torquay Museum, but the remainder of the material is distributed between various museums.

There is a great deal for the caver and visitor interested in caves to do and see in Devon and much information can be obtained from the Pengelly Research Centre, where an incredible amount of work has been done and is still being done, mostly by voluntary labour, in restoring and converting the old buildings, one into a small museum and the other into a library, lecture room and hostel.

ELEVEN

Lead Mines

MINING FOR LEAD has been responsible for the discovery of many natural caves, such as Lamb Leer, Hutton and Banwell in the Mendips, Speedwell and Hucklow Caverns in Derbyshire. Occasionally in a caving area, a chimney stack may be seen, together with some ruined buildings, all that is left of the lead resmelting industry, and around these ruins, the dimpled ground of the earlier lead miners, the 'old men'.

There was very little that the old miners did not know about limestone as the result of their continual search for lead ore, or galena veins or 'rakes' as they called them. It was often a poor living, and in 1653 Edward Manlove wrote:

> For the miners spend much money, pains and time,
> In sinking shafts before lead oar they find,
> And one in ten scarce finds, and then to pay
> One out of ten, poor miners would dismay.

The "one out of ten" was the tithe which the miners had to pay to the mineral owner for the right to work the ore.

As lead mines and caves are to be found in the same areas, mine exploration has become part of the cavers' activities. They are dangerous places. Rotting timbers or stemples, covered with stone rubble from the roof, may look like natural stone floors of passages and these rotten timbers may bridge hidden depths below. This may be said of many other mines as well, but the lead mines have been disused for a great many years and what were once stout, solid timbers are now wet, rotten and fungus covered.

Not all the mining developed into deep shafts. Much of it was near the surface, depending on the depth of the vein. As the early mining was done by individual prospectors or small groups, each with its own territory, a number of excavations would proceed at the same time. The humps and hollows of their workings have left their mark on the ground and a number of these areas are still known to farmers today as 'gruffy grounds'. The miners' workings were known as 'gruffs', 'grooves' or 'groves' and the miners themselves as 'groviers' or 'groovers'.

Farmers often refer to the old mining ground as 'unclean' land, because they believe the earth has been contaminated by 'flight', the old miner's term for the wind- or water-borne lead dust which emanated from the smelting hearths, although the last of the old resmelting works was closed during the first years of this century and much of the gruffy ground dates back to the seventeenth century.

The Vicar of Frome, the Reverend Joseph Glanvil who wrote in 1668, tells us:

There is a *flight* in the smoak, which falling upon the Grass, poysons those Cattel that eat of it. They find the taste of it upon their lips to be sweet, when the smoak chances to fly in their Faces. Brought home and laid in their houses, it kills Rats and Mice. If this flight mix with water, in which the Oar is wash't, and be carried away into a streame, it hath poisoned such Cattel, as have drunk of it after a Current of 3 miles. What of this *flight* falls upon the sand, they gather up to melt in a Slagghearth, and make Shot and Sheet-lead of it.

To me, the seventeenth century is the most interesting period of lead mining. Then, the gruffy grounds must have presented hives of activity. Numerous windlass frames known as 'stillings' or 'stillions' on Mendip or 'stowes' in Derbyshire would have marked the workings. There would have been the sound of groovers, laboriously hacking away at the rock, and the creaking of the windlasses as the laden elm buckets were drawn to the surface. The now grass-covered spoil heaps would have been grey with limestone debris and dotted about the area would be the stores or tool sheds of the miners.

The miners were not owners but licensees. Anyone who had not been 'banished' from the area could obtain a licence to work what he considered was a suitable piece of ground, unless it had already been licensed. The licence was granted by the mineral owner or the lead reeve on his behalf who generally administered the estate.

A mining area was subject to its own laws and penalties, and although basically the same, there were regional variations. T. Cox in his *Magna Britannia: Somersetshire* of 1727, when referring to the Mendip mines, stated that when a miner was found guilty of theft of tools or ore from another miner,

> He is shut up in one of their Hutts; and dry Fern, Furzes, and some other combustible Matter being set round is set on Fire. When it is on Fire the Criminal, who has his Hands and Feet at Liberty is allowed ... (if he can) to break down the Hutt, and having made himself a Passage get free, and be gone; but he must never more come to work among them more ever have any more to do about the Hills. This they call The Burning of the Hill.

In Derbyshire there was a different penalty. Edward Manlove, formerly steward of the Barghmoot Court, incorporated the mining code into a lengthy poem printed in 1653:

> For stealing oar twice from the minery,
> The thief that's taken fined twice shall be,
> But the third time that he comits such theft,
> Shall have a knife struck through his hand to th'haft,
> Into the stow, and there till death shall stand,
> Or loose himself by cutting loose his hand;
> And shall forswear the franchise of the mine,
> And always loose his freedom from that time.

When a miner had received his licence, the extent of his territory had to be determined. To do this he dug a hole waist deep from which he threw the 'law-hack' in opposite directions in line with the vein and where it fell were the limits of his claim. A hack was the miner's digging tool. but to avoid unfair advantage due to lightness in weight or the way in which it was thrown, a hack was kept especially for defining the boundaries, known as the 'law-hack', and it had to be thrown in a prescribed manner. Joseph Glanvil wrote:

> The *Groove* is 4 foot long, 2½ foot broad, till they meet a stone when they carry it as they can. The *Groove* is supported by Timber of a Divers bigness, as the place gives leave. A piece of an Armes bigness will support 10 tun of Earth ... If they cannot cut the Rock, they use Fire to aneale it, laying on Wood and

Coale and the Fire so contriv'd, that they leave the Mine before the Operation begins, and find it dangerous to enter again, before it be quite clear'd of the Smoak; which hath killed some . . . They convey out their Materials in Elme-buckets drawn by Ropes. The Buckets hold about a Gallon. Their Ladders are of Ropes . . . They beat the Oar with an iron flat piece; cleanse it in Water from the dirt; sift it through a Wire-sive. The Oar tends to the bottom and the Refuse lies at the top.

Fire and water for breaking up rock in mines later gave way to explosives of which the earliest known use in this country was by German miners in Staffordshire.

The boundaries of a groove were marked by pieces of wood known as 'draught stillings' or 'stillions'. The miner would dig his pit until he reached a vein or 'rake', which could go in any direction according to the lie and fissures of the rock. Following the rake, he would dig out underground passages, sometimes breaking into natural cave systems. The entrance shaft was usually kept narrow to avoid unnecessary work and the groover reached the vein level by using the 'back and foot' method, i.e. his back and hands against one wall and his feet against the opposite face, or he would climb up and down by a knotted rope. When he got the fragments of the vein to the surface, he would pound them to make them smaller and suitable for washing.

Often there was no water in the immediate vicinity of the mine and then the ore had to be carted to the nearest washing place. In some areas in the sixteenth century, the local lord, who was the mineral owner, provided common washing and smelting buildings. This system also made the collection of the tithes or 'lot-lead' much easier.

There were safeguards against a miner leaving a groove unworkéd. In Derbyshire, the stow had to be erected within three days of the granting of the licence and if, at any time, the groove was unworked for three weeks, except for reasons of flooding or high winds, a nick was made in the windlass spindle. If unworked for the next three weeks, a second nick was made. If, after a further three weeks, there were still no signs of the miner resuming the excavation, a third and final nick was made. Then the stow was taken down and the mineral owner was free to offer the groove to another applicant.

Relationships in the lead-mining community did not always go smoothly. There were a number of incidents which led to bloodshed and on the whole they were a rough lot, although naturally there were exceptions. Edward Manlove tells us in his poem that if any blood was shed or "any tumults raised" or other men's stows broken or defaced or any miner brought weapons to the site, they would be fined by the minery court. Vandalism and the carrying of offensive weapons seem to have been as much of a problem in the seventeenth century as they are today. It was not always the 'groovers' who caused trouble. There were often marauding parties, which sometimes included even members of the gentry, looking for trouble and loot.

Bert Russell, who helped his father during the early part of this century to remove the machinery from the nineteenth century resmelting works at St Cuthbert's at Priddy, Mendip, told me that the mining laws had not been repealed. Whether this is the case or not, they are unlikely to be administered today, since apart from the brutal penalties, the groovers and their courts have long since gone. They were not laws of State, but local mining codes. Indeed, on more than one occasion the State tried to intervene, but without success, although on some occasions an appeal was made to the State to adjudicate on a particular issue.

The ore was washed clean of earth and other matrix in troughs known as 'buddles' which were usually filled from a stream by means of 'launders'—open wooden conduits. The washing was done through sieves, as was the gold washing of the old prospectors.

When it was washed, the ore was ready for smelting and the method by which this was done seems to have varied between the different mining areas. In Derbyshire, an early form of smelting was by 'boles'. There was no mechanical aid, the necessary blast being provided by the wind. The boles were stone conical buildings, with holes round the base to catch the prevailing wind and vent holes at the top. As one might imagine, they were placed in the most windy places. These primitive blast furnaces were not peculiar to

Derbyshire, as examples are known from all over the world. By the thirteenth century, the boles had become obsolete and an artificial blast introduced by bellows. By this method far more lead could be extracted. Water power could be used for operating the bellows, but the quantity and force of water required usually made this method impossible in lead-bearing areas.

The quality of lead varied from one mining area to another as well as within each area. Some districts produced good sheet lead and others metal only suitable for shot. The most productive lead-mining area was Derbyshire, where Peak lead was not only used for repairing church roofs within the county, but was also purchased by other lead-bearing areas, such as Mendip, for this purpose. Derbyshire enjoyed an appreciable overseas export trade and, although Bristol was an export port for lead from the neighbouring Mendips, much of the lead shipped from Bristol was from Derbyshire. In the sixteenth century "Peake Leade" was described as the best in England and Mendip lead was said to be better than that worked near Richmond in Yorkshire.

As in other industries through the ages, the prosperity of the "Mynedry occupation" varied from time to time with demand, communications, the use of obsolete processes, competition from foreign sources and other reasons. Export was sometimes affected by wars, civil and foreign. In the early seventeenth century, lead could not be exported except under licence, presumably to prevent it being used by the enemy for shot.

In Mendip, improvements in methods were often far behind those of Derbyshire, not for lack of initiative among the Mendip miners, but because the poorer ore did not justify the capital outlay for improvements. In 1825, a final blow was given to a failing industry when the duty on imported lead was greatly lowered. About 1850, some of the less productive areas, such as Mendip, were on the point of closure and areas like Derbyshire in a rapid decline. The lead industry was to have a revival for a spell later, but this was in the form of resmelting activities rather than mining.

With the decline in lead mining, interest increased in the exploitation of other materials which had previously been discarded. One of these was calamine, first exploited on Mendip and later in Derbyshire, but whereas Derbyshire lead was always considered superior to that of Mendip, the reverse was the case for calamine. Calamine mixed with copper produced brass, and some time later it was found that it could produce zinc. Apart from its use for other purposes, brass was in demand for cannon, and the calamine from Mendip, the first area in Britain to produce calamine, contributed a great deal to the brass foundry business in Bristol. This industry was to become world famous, but in turn it declined during the last part of the nineteenth century. The calamine workings produced gruffy grounds as did the earlier lead mining.

Today, in the caving areas, there is little left but the gruffy grounds to mark the activities of the old groovers. The buildings that remain are usually of later date, when companies set up establishments with improved equipment to resmelt the debris or 'slimes' left behind by the old miners. There were still one or two grooves being worked at that time, but actual mining was as good as dead because of foreign competition. The resmelters did not dig for ore. The earlier miners had not been blind to the fact that the waste material from their workings still contained quite a lot of lead and they resmelted it on what were called slag-hearths, but even then there was still a good percentage left behind. The ruins of the resmelters' buildings are still a familiar sight in the cave areas such as at Grassington Moor in Yorkshire and at St Cuthberts at Priddy in the Mendips.

It was evident that a certain amount of lead went up in smoke during smelting—the 'flight'. By constructing a series of very long horizontal flues at ground level and connecting the smelting house to a distant chimney, much of the soot was left on the walls and floors of the flues. Periodically men would go along the stone tunnel-shaped flues, shovelling up the soot and bringing it back to extract any lead it might contain. This system was introduced first in Derbyshire about 1780 and later used in other lead-mining areas. Good examples of the flues can be seen at Priddy and Charterhouse on Mendip, at Grassington Moor, Yorkshire and a number of other places. The culvert-like flues usually have barrel roofs.

Close to the old resmelting buildings were the circular

buddles, now often represented as circular depressions, distinguishable from the earlier mine pits by their completely circular appearance, their comparative shallowness and the fact that they are in rows and are larger in circumference. Originally they were lined, but as at Velvet Bottom, Charterhouse, Mendip, and other places, they are now often only seen as grass depressions.

Originally, the floors of these buddles sloped down towards the centre where a vertical pole supported one or more crosspieces from which hung sails of leather or other material. Nearby was a trough, into which the ore debris was shovelled and the trough filled with water. From the trough there was a low-level exit, guarded by a grid to prevent heavy material escaping. Carrying finer material, the water flowed through this exit, down the connecting launder and into the buddle. The sails of the buddle were revolved, often several buddles being operated at the same time, from a central capstan driven by a horse or a donkey.

As the water rose in the buddles, the sails were raised so that they skimmed the surface of the water. This caused the sediment to separate into coarser material round the poles. while the finer material was carried to the edges of the buddles. The coarser the material, the greater the mineral content and so the sediment was graded outwards from the pole as 'head', 'first and second middles' and 'tailings', that nearest the buddle perimeter being unclassified as it had no mineral content. The 'head' was removed to the smelting house, but the middles were usually returned to the trough to mix with other slimes and the process repeated. Finally, due to the rising cost of lead recovery and foreign competition, the lead resmelting industry came to an end in the early part of this century.

Many of the old inns of the caving areas were frequented by the old miners and later by resmelters, using terms unfamiliar to us today—'gruffs', 'launders', 'buddles' and 'the old men', the term for their own predecessors. Long before them, back into the earlier Middle Ages and Roman times, lead had been worked and so these men were the inheritors of an ancient industry. At weekends today, the inns are full of cavers using unfamiliar words of a different kind such as 'pots', 'pitches', 'avens' and 'sumps'.

The cavers' climbing boots resound on the same paving as did the boots of the groovers. In some inns, unfortunately, you can no longer see the flagged floors, for 'improvements' have covered them with fitted carpets. In place of the door to the back kitchen where the landlord served from the barrels of beer and cider, there is often now a bar counter with its colourful beer machines and shelves displaying a whole range of spirits and concoctions that the miners had never heard of. Small tables and comfortable chairs replace the solid wooden table and settles used even in my own early caving days. The shove-halfpenny champion would show his skill on the slate at one end of the table, while at the other end would be the Northampton skittles, not to be used at the same time for fear of interference with the course of that brass disc, so deftly pushed across the chalked slate by the shove-halfpenny player's palm.

The skittle alley seems to be coming into its own again, not as the old 'beer and skittles', but in a sophisticated way. It has now become popular as a social event and the pint mugs that monopolized the ledges of the alley have made way for gin and tonics, since women, whose presence would have horrified the lead workers, are keen participants in the game. Perhaps this interest in the skittle alley has replaced in some inns the popularity of the dart board and the well-pitted board with its whitened rim made from half a tyre is not such a familiar feature as it once was, when the Pig and Whistle team of Upper Hampton would visit the Lone Fox in Lower Hampton to play the Lone Foxes or vice versa. This team visiting was part of the publican's trade, but now that the car and coach have made many of our remote moorlands readily accessible, the tendency of townspeople to invade the country inn has, not unnaturally, turned the landlord's attention to catering for distant visitors. I know of one establishment where the subdued lighting and décor of the steak room obscures the bare walls of the old skittle alley. Not only do dart teams take a second place in some pubs, but even cavers, with their heavy boots, muddied clothes and bits of rope, are no longer welcome, for the old bar parlour has changed and dingy wallpaper and tattered announcements have given way to expensive decorations and the latest in pin-ups.

Some of the old type of inns still exist and it is a rare delight to secure a place by the roaring log fire and exchange a few words about the countryside with local people instead of tourists.

Flourishing though trade may have been during the mining and resmelting days, a number of inns were faced with closure when the lead industry declined. The memory of the mines still exists in some of the inn signs today, such as the Miners' Arms, although I cannot think that many miners aspired to such heraldic distinction.

Glossary

Abseil Method used by rock climbers to descend a rock face by using the rope round the body in such a way that they can check their progress.

Adit A horizontal or slightly inclined passage entering the mine from the exterior in contrast to a shaft.

Anemolite See *Helictite.*

Anthodite Aragonite or gypsum crystals in the form of a flower (e.g. gypsum flower).

Aragonite A type of crystalline calcium carbonate similar but not the same as *Calcite.*

Aven In British caving a chimney from a subterranean passage. In France a pot or pothole from the surface.

Breccia A deposit of naturally cemented broken and angular pieces of rock. If containing bones it is a bone-breccia.

Calcite A common form of calcium carbonate found in caves.

Cave Pearls A pearl-like form of calcite or aragonite round a nucleus. The term is usually confined to smooth specimens. Rough incompleted specimens are normally termed pisolites or pisoliths.

Clints A term used in north-east England for limestone exposed on the surface forming multiple jointed 'pavements'.

Column Pillar of rock or one formed by the joining of a stalactite and stalagmite, i.e. stalagmitic column.

Conglomerate Hard mixture of different rocks, i.e. pudding stone.

Corrasion The wearing action caused by waterborne rocks or pebbles or by waterborne grit or sand.

Curtain Stalagmitic formation in which the folds have the appearance of drapery.

Doline A surface depression above a cave system.

Drapery Stalagmite having folds similar to drapery.

Dripstone Stalagmitic deposits caused by drips, i.e. stalactites, not in wide flows as in *Flowstone.*

Duck A waterlogged passage that can be passed by ducking the head beneath the water.

Eccentric Stalagmitic formations whose shape does not conform to the usual pattern. (*Helictites, Heligmites.*)

Erratic In caving an eccentric formation, but geologically it is a rock brought by natural agency, i.e. glacier ice, and deposited some distance from its origin.

Fault A crack in rock due to earth movement and pressures, etc., and causing some movement between opposite faces.

Flowstone A continuous deposit of stalagmite over floor or wall areas caused by a flow of water and not by individual drops as with *Dripstone*.

Gour A French word used in English cave terms as an alternative to 'rimstone pool'.

Grike Vertical fissures in clints.

Ground Water Water at and below the water table.

Gypsum Crystalline hydrated calcium sulphate formed by the effects of sulphuric acid which may be produced by the decomposition of iron pyrites (fool's gold), on calcium carbonate.

Gypsum Flower Gypsum crystals producing a flower form.

Haematite An iron ore of iron oxide, dark red in colour.

Headers The sides of a pliable ladder extending above the top rung which are roped by tethers to a belay or connected with the bottom extensions or 'tails' of another ladder section to lengthen it.

Helictite A distorted stalactite.

Heligmite A distorted stalagmite.

Karabiner An oval or egg-shaped steel link incorporating a spring clip used by cragsmen and cavers often attached round the waist for swiftly clipping round a rope.

Karst The general geological features above and below ground of cave areas—a German word originally applied to the Karst of Yugoslavia but now used to describe any limestone cave area—karst or karstic phenomena.

Keld A Yorkshire term for a large resurgence.

Lifeline A rope attached to a caver or cragsman independent of a climbing rope intended to take the climber's weight in an emergency.

Limestone Sedimentary rock consisting predominantly of calcium carbonate and varying from soft limestone such as chalk to hard caving limestone such as that of coal measures, i.e. Carboniferous Limestone.

Limonite An iron ore of hydrated iron oxide, varying from yellow to brown.

Living Cave The part of a cave still being formed by water action.

Master Joint A major joint.

Maypole Metal tubular pole made in sections to which a pliable ladder is attached and raised to enable climber to reach high level passage.

Moon Rock or Mountain Milk Calcium carbonate of a powdery or spongy nature either as a superficial deposit or in the form of stalactites.

Neolithic Appertaining to the New Stone Age.

Network An intricate system of passages.

Ogof Welsh word for 'cave'.

Ox Bow A loop passage joining a stream passage, formerly the original course of the stream.

Palaeontology The study of fossils.

Palaeolithic Appertaining to the Old Stone Age.

Permeability The degree to which water can pass through rock either due to the porosity of the rock itself—primary permeability—or by means of cracks, joints—secondary permeability.

Phreatic Phreatic action is the solution of rock below the water table by phreatic water, i.e. water of the phreatic zone—the saturated zone below the water table.

Pillar A tall, thin stalagmite which does not touch the ceiling.

Pisolith The cave pearl.

Pitch A vertical or near-vertical shaft.

Piton A steel spike driven into rock to provide a belay.

Pleistocene A geological period corresponding more or less to the Palaeolithic and includes the Ice Ages.

Pothole Strictly a vertical shaft open to the surface but in wider sense any cave containing pitches. Often shortened to 'pot' although this term is often used for a hollow in a stream passage.

Resurgence The point of exit of underground cave water (*Keld*).

Rimstone Pool A basin the borders of which are formed by crystalline deposit and are often in a series (*Gour*).

Rock Mill A hollow in the bed of a stream formed by the wearing action of stones being driven round by a circulatory movement of water.

Ruckle A northern term for a mass of boulders, but also used in the south, e.g. Boulder Ruckle in Eastwater Cave, Mendip.

Scalloping Small contiguous shallow hollows often continuous along the walls of a stream passage due to water. Direction of the ancient stream flow can often be ascertained from scalloping.

Shakehole Surface depression due to an underlying cave system.

Sinkhole or Sink The point at which water sinks into a cave system.

Slickensides The opposing faces of a fault may be deeply

scratched due to the movement of one face against the other, but sometimes the abrasive action is so great that the faces become very highly polished and are known as slickensides.

Slocker On East Mendip, term for a swallow hole or swallet, although not now in general use.

Sough A level driven in a mine to drain off water.

Speleology The science of caves.

Squeeze A cave passage through which it is only possible to pass by squeezing.

Stalactite Cave formation which hangs from the roof.

Stalacto-Stalagmite A stalagmitic column.

Stalagmite Floor formation, the converse of stalactite, although also the general word for all formations.

Straw A thin, hollow, straw-like stalactite.

Sump or Trap Submerged passage connecting air-filled passages at either end.

Swallet—Swallow—Swallow Hole Depression where water enters a cave system. (Sink Hole and Slocker).

Syphon Strictly a submerged passage like a plumber's U-trap, where under certain conditions water syphons from one level to another. Formerly, the term was in general use to incorrectly describe sump, duck or trap. 'Syphon' now only used to describe true syphons.

Tails The extension of the sides of a pliable ladder below the bottom rung (converse *Headers*).

Threshold The area within the mouth of a cave into which daylight penetrates.

Trap See *Sump*.

Troglobite Animal which lives only in the dark zone of the cave and leaves it only by accident.

Troglophile Animal which lives in the dark zone, but also found elsewhere.

Trogloxene Animal not resident in but a visitor to the dark zone.

Vadose The vadose zone is the rock above the water table as opposed to *Phreatic*. Vadose action is the action of water percolating between the surface and the water table.

Water Table The dividing line between the *Phreatic* and *Vadose* zones, i.e. the surface level of the saturated zone.

List of British Caving Clubs

Derbyshire Caving Club
Derbyshire Pennine Club
Eldon Pothole Club
Peakland Archaeological Society

Yorkshire

Bewerley Park Centre for Outdoor Pursuits
Bradford Pothole Club
Burnley Caving Club
Craven Pothole Club
Hollowford Training Centre
Ingleborough Community Centre
National Scout Caving Activity Centre
Northern Pennine Club
Northern Speleological Group
White Rose Pothole Club
Yorkshire Subterranean Society

SOUTH

London

Chelsea Speleological Society
Croydon Caving Club
Westminster Speleology Group

Kent

Cantium Cave and Mine Research Group

Royal Forest of Dean

Royal Forest of Dean Caving Club

Devon

Devon Speleological Society
Plymouth Caving Group

Mendip

Axbridge Caving Group and Archaeological Society
Bristol Exploration Society
Cerberus Speleological Society
Cotham Caving Group

Mendip Nature Research Committee
Shepton Mallet Caving Club
South Bristol Speleological Society
Torchlight Caving Club
Wessex Cave Club

WALES

Brynmawr Caving Club
Cwmbran Caving Club
Hereford Caving Club
South Wales Caving Club

OTHERS

Bowline Caving and Climbing Club
Furness Underground Group
North Cheshire Caving Club
Salisbury Caving Group
Severn Valley Caving Club
Swindon Spelaeological Society

Bibliography

GENERAL

Britain Underground by N. Thornber (Dalesman, Clapham, Yorks, 1953)
British Caves and Potholes by P. R. Deakin and D. W. Gill (1975)
British Caving edited by C. H. D. Cullingford (Routledge & Kegan Paul, 1962)
Cave Craft An introduction to caving and potholing by D. Cons (Harrap and Co., 1966)
Cave Hunting by W. Boyd Dawkins (1874; republished by E. P. Publishing Ltd, 1973)
Caves by Tony Waltham (Macmillan, 1974)
Caves and Caving A Little Guide in Colour by Marc Jasinski (Paul Hamlyn, 1967)
Caving by E. A. Baker (Chapman & Hall, 1932; republished by E. P. Publishing Ltd, 1970)
Caving and Potholing by D. Robinson & A. Greenbank (Constable, 1964)
Challenge Underground by Bruce L. Bedford (1975)
Exploring Caves by C. H. D. Cullingford (Oxford University Press, 1951)
First Book of Caves by Elizabeth Hamilton (Edmund Ward, London, 1964)
Irlande et Cavernes Anglaises by E. A. Martel (Paris, 1897)
Life and Death Underground by J. Lovelock (Bell & Sons, 1963)
Manual of Caving Techniques edited by C. H. D. Cullingford (Routledge & Kegan Paul)
Reliquiae Diluvianae by W. Buckland (1823)
Science of Speleology (British Cave Research Association)
Tour to the Caves by John Hutton (E. P. Publishing Ltd.)

SHOW CAVES

Discovering caves, A Guide to Show Caves of Britain by Tony and Anne Oldham (Shire Publications, 1972)

MENDIP

The Complete Caves of Mendip by N. Barrington and W. I. Stanton (Cheddar Valley Press, 1970)

History of Mendip Caving by Peter Johnson (David and Charles, 1967)

Lamb Leer (2 parts) (Mendip Nature Research Committee)

Limestone and Caves of the Mendip Hills edited by D. I. Smith and D. P. Drew (David and Charles)

Mendip—Cheddar its Gorge and Caves by H. E. Balch (John Wright, 1947)

Mendip—The Great Cave of Wookey Hole by H. E. Balch (Clare, Son & Co., Wells, 1929)

Mendip—Its Swallets, Caves and Rock Shelters by H. E. Balch (John Wright, Bristol, 1948)

The Mendips by A. W. Coysh, E. J. Mason, V. Waite (Robert Hale, 1971) (Chapter V)

The Netherworld of Mendip by E. A. Baker and H. E. Balch (J. Baker & Son, 1907)

Pioneer under the Mendips, A Biography of H. E. Balch by W. I. Stanton (Wessex Cave Club, 1969)

Story of Wookey Hole by E. J. Mason (Wookey Hole Caves, 1970)

Wookey Hole, its Caves and Cave Dwellers by H. E. Balch (Oxford, 1914)

WALES

Caves in Wales and the Marches by D. W. Jenkins and A. Mason Williams (Dalesman, Yorks, 1963)

Goat's Hole, Paviland Gower by F. J. North, *Annals of Science V No. 2*, pp. 91–128 (1942)

Gower Caves by J. G. Rutter and E. E. Allen (Welsh Guides, Swansea, 1948)

Ogof yr Esgyrn by R. H. D'Elboux, *Archaeologia Cambrensis*, p. 113 (1924)

Ogof yr Esgyrn E. J. Mason, *Archaeologia Cambrensis*, pp. 18–71 (1968)

Portrait of the Brecon Beacons by E. J. Mason (Robert Hale, 1975) (Chapter 10)

Prehistoric Gower by J. G. Rutter (Welsh Guides, Swansea, 1948)

DEVON

The Caves of Devon by A. D. & J. E. A. Oldham and J. Smart (published privately by T. Oldham)

SCOTLAND

The Caves of Scotland (except Assynt) by Tony Oldham (published privately by T. Oldham, 1975)

NORTHERN CAVES

Cave Hunting Holidays in Peakland by G. H. Wilson (Chesterfield, 1937)
Caves and Caverns of Peakland by C. Porteus (Derby, 1950)
Caves of Derbyshire by T. D. Ford (1967)
Gaping Gill and Ingleborough Cave (Bradford Pothole Club)
Limestones and Caves of North West England by A. C. Waltham and M. M. Sweeting (David and Charles, 1974)
Northern Caves by various authors (Several volumes) (Dalesman Books)
Pennine Underground by N. Thornber (Dalesman, Clapham, Yorks, 1965)
Potholing Beneath the Northern Pennines by D. Heap (1964)
Some Caves and Crags of Peakland by G. H. Wilson (1934)
Yorkshire Caves and Potholes by A. Mitchell. No. 1 North Ribblesdale (1945), No. 2 Ingleborough

CAVE DIVING

Caves and Cave Diving by G. de Lavaur (Robert Hale, 1956)
The Mendips by A. W. Coysh, E. J. Mason, V. Waite (Robert Hale, 1971) (Chapter IV)

MINING

Derbyshire's Old Lead Mines and Miners by J. H. Rieuwerts (Moorland Publishing Company, 1972)
Lead Mining in the Peak District by T. D. Ford and J. H. Rieuwerts (Peak Park Planning Board, 1970)
Mines of Mendip by J. W. Gough (Oxford, 1930; republished by David and Charles, 1967)

PERIODICALS AND PROCEEDINGS

Archaeologia Cambrensis (various)
British Association reports (various)
British Caver, edited and published by Tony Oldham
Bulletins of the Peak District Mines Historical Society
Caves and Caving, journal of the British Speleological Association (discontinued)

Descent—a *Magazine for Caves*, Descent Publications Ltd, Wells
Geographical Journal (various)
Journal of the Derbyshire Archaeological and Natural History Society (various)
Journal of Geology (various)
Journal of the Yorkshire Ramblers Club
Proceedings of the Devon Archaeological Exploration Society (various)
Proceedings of the Geological Association (various)
Proceedings of the Prehistoric Society (various)
Proceedings of the Somerset Archaeological and Natural History Society, Taunton (various)
Quarterly *Journal* of the Geological Society (various)
The Speleologist (discontinued)
Studies in Speleology, Association of the Pengelly Cave Research Centre
Transactions and *Bulletins* of the British Cave Research Association
Transactions of the Carmarthen Antiquarian Society (various)
Transactions of the Cave Research Group of Great Britain (discontinued)
Transactions of the Torquay Natural History Society (various)
Wells Natural History and Archaeological Society's Annual Reports, Wells (various)

FILMSTRIP/SLIDE SETS AND NOTES

Caves: origins, development and formations
Caving and Potholing Techniques
Limestone Landforms,
    All by Alan C. Coase (Diana Wyllie Ltd)

Index